WASTE MANAGEMENT AND SUSTAINABLE CONSUMPTION

The accelerated pace of global consumption over the past decades has meant that governments across the world are now faced with significant challenges in dealing with the dramatically increased volume of waste.

While research on waste management has previously focused on finding technological solutions to the problem, this book uniquely examines the social and cultural views of waste, shedding new light on the topic by emphasising the consumer perspective throughout. Drawing on a wide variety of disciplines including environmental, economic, social and cultural theories, the book presents philosophical reflections, practical examples and potential solutions to the problem of increasing waste. It analyzes and compares case studies from countries such as Sweden, Japan, the USA, India, Nigeria and Qatar, bringing out valuable insights for the international community and generating a critical discussion on how we can move towards a more sustainable society.

This book will be of great interest to postgraduate students and researchers in environmental policy, waste management, social marketing and consumer behaviour, as well as policymakers and practitioners in consumer issues and business.

Karin M. Ekström is Professor of Marketing at the University of Borås, Sweden. Her research interests are the meaning(s) of consumption, sustainable consumption, consumers' relations with artefacts, collecting, consumer socialization and family consumption. Current research projects involve culinary tourism and market orientation of art museums.

"In a global economy dependent upon ever rising levels of consumption, waste is the material artefact that is steadily consuming fragile eco-systems and accelerating the pace of climate change. In this book, an interdisciplinary group of scholars analyze waste not merely as a technical problem but as configurations of practices and meanings that are embedded in broader socio-cultural systems and institutional structures. This theoretically innovative, cross cultural and eminently practical study of waste, and how societies can better manage its deleterious consequences, is both timely and fascinating."

Craig J. Thompson, University of Wisconsin–Madison, USA

"In gathering more contributors to 'waste's table', this volume demonstrates the claim made by Richard Wilk in its afterword: that waste is the single most important link between population, consumption and climate change. In showing how waste is inexorably a matter of consumption, it will surely gather more contributors to the debate."

Nicky Gregson, Durham University, UK

"This volume has an absolutely essential message: achieving real sustainability requires understanding the social and cultural dimensions of consumption. In a carefully composed set of nuanced powerful chapters, some of the most eminent scholars in the field of critical consumption studies provide a wealth of arguments that help us realize why people consume and why the vicious cycles of production and consumption are so extremely hard to break. Here we learn, once and for all, that this concerns the habits of the heart and the deepest layers of identity production and that sustainable consumption simply will not be achieved just by adding more sophisticated recycling technologies or incentives to conserve."

Sverker Sörlin, KTH Royal Institute of Technology, Sweden

WASTE MANAGEMENT AND SUSTAINABLE CONSUMPTION

Reflections on consumer waste

Edited by Karin M. Ekström

LONDON AND NEW YORK

First published 2015
by Routledge
2 Park Square, Milton Park, Abingdon, Oxon OX14 4RN

and by Routledge
711 Third Avenue, New York, NY 10017

Routledge is an imprint of the Taylor & Francis Group, an informa business

© 2015 Karin M. Ekström

The right of the editor to be identified as the author of the editorial material, and of the authors for their individual chapters, has been asserted in accordance with sections 77 and 78 of the Copyright, Designs and Patents Act 1988.

All rights reserved. No part of this book may be reprinted or reproduced or utilised in any form or by any electronic, mechanical, or other means, now known or hereafter invented, including photocopying and recording, or in any information storage or retrieval system, without permission in writing from the publishers.

Trademark notice: Product or corporate names may be trademarks or registered trademarks, and are used only for identification and explanation without intent to infringe.

British Library Cataloguing-in-Publication Data
A catalogue record for this book is available from the British Library

Library of Congress Cataloging-in-Publication Data
Waste management and sustainable consumption: reflections on consumer waste / edited by Karin M. Ekström.
pages cm
Includes bibliographical references.
1. Source reduction (Waste management) 2. Waste minimization.
3. Refuse and refuse disposal—Social aspects. I. Ekström, Karin M., editor.
TD793.95.W37 2015
363.7—dc23   2014026986

ISBN: 978-1-138-79725-3 (hbk)
ISBN: 978-1-138-79726-0 (pbk)
ISBN: 978-1-315-75726-1 (ebk)

Typeset in Bembo
by Swales & Willis Ltd, Exeter, Devon, UK

Printed and bound in the United States of America by Publishers Graphics, LLC on sustainably sourced paper.

CONTENTS

List of figures and tables   vii
Contributors   ix

    Introduction   1
    Karin M. Ekström

PART I
Consumption and waste   11

  1 Recycling the home: the constant flow of domestic stuff, emotions and routines   13
    Orvar Löfgren

  2 The curse of the new: how the accelerating pursuit of the new is driving hyper-consumption   29
    Colin Campbell

  3 Thinking waste sociologically   52
    Paul Hewer

PART II
Managing waste   65

  4 Factors affecting development of waste management: experiences from different cultures   67
    Mohammad J. Taherzadeh and Karthik Rajendran

5   Waste prevention action nets   88
    Hervé Corvellec and Barbara Czarniawska

6   Curbside cartographies in an urban food-waste
    composting program   102
    John W. Schouten, Diane M. Martin and Jack S. Tillotson

7   Cloth Loop: an attempt to construct an actor-network   115
    Eva Gustafsson, Daniel Hjelmgren and Barbara Czarniawska

PART III
Socio-cultural views on waste   **131**

8   Exploring food waste through the lens of social practice
    theories: some reflections on eating as a compound practice   133
    Dale Southerton and Luke Yates

9   Environmental consumer socialization among Generations
    Swing and Y: a study of clothing consumption   150
    Karin M. Ekström, Daniel Hjelmgren and Nicklas Salomonson

10 Unpacking corporate sustainability: sustainable
    communication, waste and the 3Rs in a network society   166
    Pierre McDonagh and Andrea Prothero

PART IV
Preventing waste   **185**

11 Upcycling of pre-consumer waste: opportunities and
    barriers in the furniture and clothing industries   187
    Daniel Hjelmgren, Nicklas Salomonson and Karin M. Ekström

12 Post-ownership sustainability   199
    Russell Belk

13 Supplementing the conventional 3R waste hierarchy:
    considering the role of carbon rationing   214
    Maurie J. Cohen

14 Afterword: the waste that matters   225
    Richard Wilk

Index   240

FIGURES AND TABLES

Figures

4.1	MSW generation rate in 63 different countries, in kg per capita per day	68
4.2	Roadmap to zero waste: factors, possibilities and challenges	78
8.1	The destinations of surplus food produced at afternoon and evening meals	142
10.1	The process of sustainable communication	168
10.2	The inverted pyramid of sustainability	179
11.1	The jacket prototype	193
11.2	The sofa prototype	193

Tables

4.1	Different countries and their waste management activities	70
4.2	Rating of social factors affecting waste management in different countries	83
8.1	Characteristics of afternoon and evening meals at which surplus food and food waste is reported	143
8.2	Binary logistic regression of food surplus at afternoon and evening meals	145
11.1	The companies' product areas and contributions to the clothing project	191
11.2	The companies' product areas and contributions to the furniture project	192
11.3	Resource combination in the clothes project – the jacket prototype	193

11.4	Resource combination in the furniture project – the sofa prototype	194
11.5	Costs and lost/reduced values that might appear in the resource network	195

CONTRIBUTORS

Russell Belk is Kraft Foods Canada Chair in Marketing at Schulich School of Business and Distinguished Research Professor at York University, Canada. His areas of specialization include meanings of possessions, collecting, sharing, gift giving, extended self, and materialism. His work tends to be cultural, qualitative, and visual.

Colin Campbell is an Emeritus Professor of Sociology at the University of York, UK. His research interests include the sociology of religion, consumerism, cultural change, and sociological theory.

Maurie J. Cohen is Associate Professor and Director of the Program in Science, Technology, and Society at the New Jersey Institute of Technology. He is also Associate Faculty Member with the Division of Global Affairs at Rutgers University, and Associate Fellow at the Tellus Institute, USA.

Hervé Corvellec is a Professor of Business Administration at the Department of Service Management and Service Studies, Lund University, Sweden, and at the Gothenburg Research Institute (GRI), University of Gothenburg, Sweden. His field of research is business and public administration, with a research focus on infrastructure services.

Barbara Czarniawska is Professor of Management Studies at the University of Gothenburg, Sweden. She takes a feminist and constructionist perspective on organizing, recently exploring the connections between popular culture and practice of management, and the organization of the news production. She is interested in methodology, especially in techniques of fieldwork and in applications of narratology to organization studies.

Karin M. Ekström is Professor (chair) in Marketing at University of Borås, Sweden. Her research interests are the meaning(s) of consumption, sustainable consumption, consumers' relations with artefacts, collecting, consumer socialization, and family consumption. Current research projects involve culinary tourism and market orientation of art museums.

Eva Gustafsson is a Senior Lecturer at Borås University, Sweden and Associate Professor in Business Administration. Her latest project concerns the development of textile recycling and sustainability in the textile industry. She is also interested in areas such as e-services, gender roles, and risk.

Paul Hewer is a Reader in the Consumption, Markets and Society group at the University of Strathclyde, Glasgow, Scotland. His research interests lie in the area of consumption and marketplace mythologies. He dwells in Scotland with his family and is a keen cook.

Daniel Hjelmgren is Assistant Professor in Marketing at University of Borås, Sweden. His research is in the fields of business networks, management of product development and sustainable consumption. Current research projects concern the reuse and recycling of clothes.

Orvar Löfgren is Professor Emeritus of European Ethnology at Lund University, Sweden. The cultural analysis and ethnography of everyday life has been the focus of much of his research. Currently he is engaged in a project on life at home.

Pierre McDonagh is Professor of Marketing at the University of Bath, UK. He primarily focuses on the (im)possibility of sustainability within the prevailing order and oppositional counter-culture. He enjoys music, good food, football and tennis, normally with friends and family. He is married to Andy Prothero with three sons Ethan, Cal and Dylan.

Diane M. Martin is Associate Professor of Marketing at Aalto University in Helsinki, Finland. Her academic research employs ethnographic methods in examining relationships between consumers, communities, and culture.

Andrea Prothero is Associate Professor of Marketing at University College Dublin, Ireland. Her research broadly explores the area of Marketing in Society. Specific research projects have focused on, for example, advertising to children, motherhood and consumption, sustainability marketing, and sustainable consumption.

Karthik Rajendran is a PhD student at the Swedish Centre for Resource Recovery, University of Borås, Sweden, working on reactor designs, techno-economic evaluation, modelling and optimization for biofuels such as biogas and ethanol.

Nicklas Salomonson is Associate Professor in Business Administration at the School of Business and IT, University of Borås, Sweden. His research concerns areas such as interaction in service encounters, customer misbehaviour and sustainable consumption. A current project aims at clarifying the customer discourse in relation to service encounters where customers misbehave.

John W. Schouten is Professor of Marketing at Aalto University in Helsinki, Finland, and at the Center for Customer Insight at University of St. Gallen, Switzerland. His work spans areas of consumer identity, consumption communities, alternative research methodologies, and environmental and social sustainability.

Dale Southerton is Professor of Sociology and Director of the Sustainable Consumption Institute at the University of Manchester, UK. His research interests focus on consumption and sustainability, theories of social change, time use and the temporal organisation of everyday life, technologies and innovation, and taste, belonging and social distinction.

Mohammad J. Taherzadeh is Professor in Biotechnology and Director of Resource Recovery at University of Borås, Sweden. He works on biological conversion of residuals to biofuels and materials.

Jack S. Tillotson is a doctoral candidate at Aalto University in Helsinki, Finland. From 2009 to 2013, he worked as an investment consultant in financial services. Reawakening his concern for the environment, his doctoral studies are focused on the unifying force of change within business to provide balance for humanity and the planet.

Richard Wilk is Provost's Professor of Anthropology at Indiana University, USA, where he runs the Food Studies program. Much of his recent work has turned towards the global history of food and sustainable consumption.

Luke Yates is Hallsworth Research Fellow in the Department of Sociology and the Sustainable Consumption Institute at the University of Manchester, UK. His research interests revolve around the sociology of consumption and eating, political movements and cultural groupings, and the everyday organisation of living arrangements.

INTRODUCTION

Karin M. Ekström

Society has over time, and particularly during the last decades, developed towards a consumer society. We have witnessed an accelerated pace of consumption as well as an increased emphasis on symbolic consumption where consumer identity is reflected by what is consumed (e.g. Bauman 1998; Ekström 2013a). Products are thrown away, not necessarily because they are worn out, but because they do not have the right colour or style in fashion. Products are also disposed of because they cannot be repaired or because the cost of repair often exceeds the cost of purchasing a new product. The increase in consumption has also led to an increase in waste and, as a result, an increased interest in waste management.

The research on waste management has mainly focused on technology, in other words, trying to find technical solutions to the problem; for example, to recover energy from waste or to recycle waste material. The technological approach is important, in particular for finding innovative solutions, but needs to be complemented with social and cultural constitutions of waste, as suggested in this book. Dealing with waste is a complex and difficult project that requires different disciplinary perspectives. Social and cultural views will contribute to new understandings of waste. Albeit waste is often a practical problem, it is also necessary to theorize about waste. The aim of this book is to present different perspectives, theories and methods for understanding waste and to contribute to critical discussions on how we move towards a more sustainable society.

The book originated in a workshop at the University of Borås, Sweden in November 2013. The workshop had the title "Reduce, reuse, recycle – environmental and social challenges" and was funded by a grant provided by The Swedish Retail and Wholesale Development Council and the Swedish Research Council for Environment, Agricultural Sciences and Spatial Planning. The support is gratefully acknowledged. The participants in the workshop represent different academic disciplines and the intention from the very beginning was to publish a book so that the vivid discussions on waste in Borås could continue around other tables and across the world after the workshop.

The book consists of four parts: consumption and waste, managing waste, sociocultural views on waste and waste prevention. Before presenting the different chapters, some reflections are given on these different topics.

Consumption and waste

Consumption is a motor of waste production and a site for potential intervention. To understand waste, it is necessary to understand consumption in-depth and in particular the driving forces for consumption. What motivates consumers to replace a fully functioning kitchen with a new one rather than just repainting it? One driving force for consumption is material necessary consumption such as purchases of food and clothes (Hjort 2004). Another driving force is conformity (Ekström forthcoming), to be like others and to be part of society (Duesenberry 1949). Consumers influence each other and consumption has, over time, become an important social marker (Ekström and Glans 2011). Bauman (1998) argues that consumption has replaced the significance of work as a status indicator. Distinction is yet another driving force for consumption (Bourdieu 1984; Hjort 2004). It is about differentiation and to appear unique. Social boundaries decide, however, how unique one is permitted to be and the situational context determines whether something is perceived as conformity or distinction (Ekström forthcoming).

What motivates consumers to reduce their consumption, to reuse and recycle? A consumer perspective is necessary to understand how consumers relate to waste and consumption and to make it easier for consumers to act in an environmentally friendly way. We need to better understand everyday life, including habits and routines. Emotions and feelings such as anger, joy and guilt towards waste also need to be recognized. The common rationality in dealing with waste is not believed to be the single solution, even though the title of the book is waste management. Furthermore, waste cannot be studied in isolation, but needs to be considered in relation to material culture. The boundaries between people and things should not be overlooked (Miller 1998). The interdependency between individuals and artefacts are discussed by Hård af Segerstad (1957, p. 15): 'people without things are helpless, but things without people are meaningless'. Relations to objects often develop over a period of time. Miller (1997) calls a gradual transformation process 'to appropriate' when discussing a study on council-flat kitchens where he found how people had tried to make a home. Also Belk (1988) discusses how an object becomes part of the self when an individual appropriates the object. He claims that possessions are not only part of the self, but can be seen as instrumental for development of the self. Relations to artefacts are discovered, in particular during processes of moving when there is an overflow of stuff, decisions, emotions and worries (Ekström 2013b). The difficulties involved in separating from things need to be recognized. Gregson (2011) concludes in an ethnographic study that identities, social relations and homes are made by both acquisition and disposal of things. The study indicates that to understand waste, an understanding of the entire consumption process is required, such as desiring, purchasing, owning, using, maintaining, repairing, reusing, redesigning, recycling and disposing. A variety of disciplines will contribute to new perspectives on consumption and waste, encourage new creative ideas and also make it possible for us to rethink consumption and waste.

Managing waste

Parallel to the increase in consumption and the resulting increase in waste, there has been a growing environmental awareness of how to manage waste. Landfills and ocean dumping were for a long time considered a solution and contributed to making waste invisible. Even though the use of landfills and ocean dumping has decreased, they still exist in many countries and cause environmental problems. Apart from environmental reasons, an increased interest in waste has arisen as a result of recognizing that waste can be turned into resources. An example is the recovery of energy from waste such as incineration plants burning waste to recover energy that is released in the process. However, potentially valuable materials are lost, as well as the energy associated with their initial production. Another example of turning waste into resources is research on recycling that attempts to find solutions on how to remanufacture raw material. For example, unravelling and re-spinning textiles into new fibres that can reduce the use of virgin raw material. An example is making fleece from recycled polyester.

A concern over environmental issues has, over time, also been recognized in the political agenda. The Brundtland report, *Our Common Future*, published by the World Commission on Environment and Development (1987) emphasized environmental issues and sustainable development. Environmental concerns, also involving waste, were highlighted during the Rio Conference in 1992, the United Nations Conference on Environment and Development. The European Union waste directive (The European Parliament and the Council of the European Union 2008/98/EC) suggests that waste should be organized according to a hierarchy, the preferred one listed first at the top and the least desirable at the bottom: reducing, reusing, recycling, incineration with energy recovery and landfilling.

It is no good if reuse or recycling is seen as a way to lessen a bad conscience for overconsumption. Reducing waste can involve reducing consumption, but not necessarily if it involves redirecting resources to be used in other ways. Reducing consumption is for economic reasons not popular in the public or political debate since the Gross Domestic Product is dependent on consumption. In particular, during economic crises, there is a risk that consumer policy issues and sustainability are downplayed. Jackson (2009) advocates prosperity without growth and emphasizes that prosperity goes beyond material wealth. Economic growth, beyond a certain point, does not seem to advance human happiness; it may even impede it (Jackson 2009). He argues that consumerism promotes unproductive status competition and that possibilities for people to flourish in less materialistic ways must be established. To reduce, however, may not always be easy, depending on the socio-cultural context, since consumption has become an important social marker. In today's society, everybody participates in the 'catwalk of consumption' and to not take part involves psychological and social risk taking (Ekström and Hjort 2009). Therefore, it is also important not to moralize against consumption (Ekström and Salomonson 2014).

Socio-cultural views on waste

A lot of research on recycling has focused on the individual, often from a psychological perspective (e.g. Guagnano et al. 1995) and there is a need to better understand the socio-cultural context. Evans et al. (2013) argue that food waste constitutes an embedded part of the economy as well as of people's relationships, identities and actions. In what situations and to what extent do people and family members influence each other regarding consumption and waste? Parents sometimes learn from their children about environmental behaviour such as recycling (e.g. Ekström 2007).

The perception of waste varies for different individuals – one man's trash is another man's treasure. Poverty and increasing inequality across the globe should also be considered when studying waste, including the role of scavengers. The perception of waste varies in different contexts, for example, what type of waste is considered good or bad and in which situations? How is the smell and touch of waste perceived in different situations? Why do activities of waste management often take place out of public sight (Wilk 2004)? The ethics of waste should be recognized (Hawkins 2006).

It is necessary to recognize that everyday life is full of mundane consumption decisions and that dealing with waste can be one of them. To deal with waste takes time and consumers must consider it meaningful and convenient to recycle to be motivated to do so; for example, recycling stations must be close to residential areas. To build a waste management strategy that is based on separating waste at the source, i.e. at the consumer's home, is today common in many Western countries. However, it requires storage space for items to be reused or recycled. Many of today's new-built houses and apartments have an appealing architecture with large windows and few walls, but unfortunately little space for storage. Also, the environmental effect of driving to the recycling station has to be considered and compared to collecting recyclable items from the home.

Consumers' awareness and willingness to reuse, recycle, and repair must be understood in relation to the socio-cultural contexts. Communication to create increased awareness must take its origin in the consumer's reality. Social media such as Facebook, blogs and Twitter may in some cases be more efficient than a brochure and also lead to less waste. The fashion industry has clearly benefited from social media and the many fashion blogs that exist. Hopefully, communication about overconsumption and waste can also become even more interesting topics of conversation in social media. The elusive and transient nature of consumption (Ekström and Brembeck 2004) makes it possible to perceive sustainable consumption as an oxymoron.

Earlier generations oriented themselves differently to consumption and waste compared to consumers of today. Their behaviour may have resembled environmental consciousness, but instead it was often a result of living with a scarcity of resources. Things were bought to be kept, maintained and repaired. There was no room for the throwaway society that the last decades of overproduction and overconsumption have encouraged. Another reflection is that some decades ago second-hand clothes were not socially acceptable among many consumers, but today they have gained widespread popularity, in particular vintage fashion. It shows that social norms can change over time. Buying second-hand items is today visible in many forms, for example, car boot sales, charity shops, retro shops and auction sales (Gregson and Crewe 2003).

Waste prevention

Even though the EU hierarchy thinks waste should be reduced, the best alternative would, however, be if waste did not happen at all. How can waste be prevented? We need to consider the early stages of consumption, before things are turned into waste. What is driving people to throw things away? What role do fashion cycles, minimalist trends in interior design, etc. play? What roles do realtors play in encouraging people to restyle their homes before sale so that they look almost empty? Will homes that earlier were overstuffed and now are suffering from symptoms similar to bulimia necessarily lead to a better quality of life? And if so, is it a better life for the individual, the society or the planet? Maybe the answer lies in how the waste is managed. Furthermore, preventing waste requires producers as well as consumers to rethink how resources are spent. Overall, there is a need for a more nuanced understanding of material flows in societies that aim towards goals such as zero waste, a circular economy, cradle-to-cradle design or closing the loop.

Recycling often concerns post-consumer waste, but it is here also important to recognize pre-consumer waste. For example, increased fashion cycles may result in an oversupply of textiles of a colour or material that has become out of fashion. Unsold products of different kinds also become waste due to a lack of demand. How can this type of waste be prevented? To reduce stock could be one strategy, even though it does not deal with the problem at cause. Redesign and upcycling are other approaches and here it is important that creativity is encouraged, for example, by encouraging a design process involving people with different areas of expertise and competence. Similar to the synergy that is often established in interdisciplinary research, this could also be encouraged to happen in a design process. Nevertheless, it is necessary to critically assess the prevalence of fast-fashion and to consider its consequences on the environment.

When discussing waste, the issue of whose responsibility it is to solve the problem is often on the table. Is it the producers, the consumers or society? The opinions differ and vary across people and cultures. Sometimes, the producers take responsibility by paying a fee in connection to packaging that covers the handling of recycling the packaging. Sometimes, the consumers pay for the waste to be disposed of by combustion and are encouraged to reduce the waste in the dustbins by the waste that is recyclable to the recycling stations. It is not realistic, however, to consider consumers as the sole agents for reducing waste and increasing recycling. Governments and companies also need to contribute to reduce waste and to move towards a more sustainable society.

Consumption and waste should not merely be looked at from a micro perspective, i.e. how individuals or companies deal with it. Meso and macro levels are also necessary by considering community and society. In particular, in a society with increased emphasis on individualism, there is a need to look at the problem of waste from a collective perspective. A few individuals who ignore the waste problem may cause harm for other individuals. Laws and taxes are necessary at times. Jackson (2009) suggests that creating resilient social communities is important. He writes:

> To reassert the crucial importance of shared endeavor is not to demonize individual needs or personal dreams. The point is to redress the balance between the self and society – in a way that re-establishes the importance of public goods in working for the benefit of us all.
>
> *(Jackson 2009, p. 203)*

To reduce consumption and waste is a challenging task, but undeniably of crucial importance for our environment. It is a collective project and not entirely dependent on the individual. It is about saving our shared future!

The structure of the book

This book consists of four parts and the contributors represent a variety of disciplines.

Part I: Consumption and waste

In the first part, researchers in ethnology (cultural anthropology), sociology and marketing present different views on consumption and waste. The first chapter, *Recycling the home: the constant flow of domestic stuff, emotions and routines*, written by Orvar Löfgren, deals with overflows in domestic life and the ways in which sustainable living is affected by the entanglement of consumer goods, activities, feelings and the dreams of a perfect home. He looks at commodities moving in time and space inside the home and how people navigate in the process of handling 'too much'.

In the second chapter, *The curse of the new: how the accelerating pursuit of the new is driving hyper-consumption*, Colin Campbell argues that it is the high value attached to the new and the novel that is the major source of hyper-consumption. Three different forms of new are identified – the new as the fresh, the innovative and the novel. The reasons for an accelerating pursuit of the new are found in both the rapidity of technological advance and the decline in the average cost of products.

The third chapter, *Thinking waste sociologically,* written by Paul Hewer, should be read as an entrée into rethinking waste through the lens of theories of consumption. Waste is reclaimed as everyday practice as expressed in the shift from order-building as farming or mining to that of gardening where the work of recycling and reassembling oneself and our relations with others and things is best understood.

Part II: Managing waste

In the second part of the book on managing waste, researchers in technology, management and marketing present different views on managing waste. In Chapter 4, *Factors affecting development of waste management: experiences from different cultures,* Mohammad Taherzadeh and Karthik Rajendran describe and compare the waste management in ten countries representing a wide variety in waste-management practices. Different factors affecting waste management are discussed. It is argued that the choice of a successful waste-management practice is a local solution and the key is not universal.

In Chapter 5, *Waste prevention action nets,* Hervé Corvellec and Barbara Czarniawska claim that it is unclear what constitutes waste prevention. A comparative analysis of three waste prevention examples is presented to address this lack of clarity: a waste-management company selling waste-prevention services to its waste-producing customers, the opportunity for Swedish householders to opt out of unaddressed promotional material and a car-sharing programme. The analysis is informed by an action-net perspective. It is concluded that waste prevention rests on the invention of new modes and patterns of interactions that both build and disrupt the existing institutional order of consumption.

Chapter 6 , *Curbside cartographies in an urban food-waste composting program,* written by John W. Schouten, Diane M. Martin and Jack S. Tillotson, takes an actor-network theory (ANT) approach to examining the creation, implementation and aftermath of a programme of residential kerbside collection of food waste for composting in a mid-sized US city. Rather than follow popular notions that cultural change begins with cognition, values and attitudes, the authors find that the problem may lie in prioritizing consumer agency over that of other actors. Key elements of the apparent success of the programme are identified and it is discussed how both human and non-human actors contributed.

In Chapter 7, *Cloth Loop: an attempt to construct an actor-network,* Eva Gustafsson, Daniel Hjelmgren and Barbara Czarniawska describe a project in which a new approach for collecting used clothes and textiles was to be tested. It was started by a well-known Swedish shopping mall in cooperation with one of Sweden's larger humanitarian organizations. The Callon-inspired analysis reveals different obstacles in the enrolment and mobilization of allies and donors that need to be dealt with to improve the inflow of used clothes and textiles.

Part III: Socio-cultural views on waste

In the third part of the book, researchers in sociology and marketing present different socio-cultural views on waste. Chapter 8, *Exploring food waste through the lens of social practice theories: some reflections on eating as a compound practice*, written by Dale Southerton and Luke Yates, locates food waste in the constituent activities of eating as a social practice, using a recent survey of eating patterns in the UK to illustrate the conceptual approach. They found that the production of leftovers and waste at meals are strongly associated with aspects of meal occasions such as provisioning and preparation, duration, meal type, and the presence of companions.

In Chapter 9, *Environmental consumer socialization among Generations Swing and Y: a study of clothing consumption*, Karin M. Ekström, Daniel Hjelmgren and Nicklas Salomonson investigate how Generation Swing (born 1930–1945) and Generation Y (born 1976–1994) experience and approach the consumption of clothing. The study is based on focus group interviews. Generation Swing shows a more restricted approach towards the consumption of clothing than Generation Y. Even though primary consumer socialization is critical, the study illustrates that through secondary socialization it is possible to influence people to adopt more environmentally conscious consumption behaviour regarding clothing.

In Chapter 10, *Unpacking corporate sustainability: sustainable communication, waste and the 3Rs in a network society*, Pierre McDonagh and Andrea Prothero advance the argument that elements of the theory of sustainable communication are at play in contemporary society and several corporate examples are considered. A version of liberal pluralism akin to corporate social responsibility could be seen more than the more radical notion of sustainable communication. Even though corporations are engaging with consumers to reduce, reuse and recycle, the current strategies are engaging with the disposal of waste at a systemic level and more work in this area is still required.

Part IV: Preventing waste

In the fourth part of the book, researchers in marketing and sustainability studies discuss waste prevention. In Chapter 11, *Upcycling of pre-consumer waste: opportunities and barriers in the furniture and clothing industries*, Daniel Hjelmgren, Nicklas Salomonson and Karin M. Ekström discuss opportunities and structural barriers of utilizing pre-consumer waste materials in the Swedish furniture and clothing industries. In-depth interviews have been conducted in a project that aims to make use of textile waste. The results show that the possibility for making use of waste materials is greater in the furniture industry than in the clothing industry due to more local and flexible production in the furniture industry.

In Chapter 12, *Post-ownership sustainability*, Russell Belk argues that owning consumer goods is not the only way to access the goods we need. Sharing and collaborative consumption are rapidly growing alternatives to individual ownership that lessen the environmental impacts of consumption. The chapter discusses different forms of post-ownership and examples of current sharing and collaborative consumption alternatives and assesses whether we may be heading towards a post-ownership society.

In Chapter 13, *Supplementing the conventional 3R waste hierarchy: considering the role of carbon rationing*, Maurie Cohen suggests that we may need to supplement the '3R' waste hierarchy with a fourth element – ration – as a result of increased attentiveness to climate change. The current status of rationing as a response to contemporary sustainability challenges are reviewed and the focus is primarily on how this concept has been taken up to date in the UK. He discusses the circumstances that would need to become manifest for a relatively stringent programme of consumer regulation around carbon to become politically viable.

References

Bauman, Zygmunt (1998), *Work, Consumerism and the New Poor*. Buckingham: Open University Press.
Belk, Russell W. (1988), 'Possessions and the extended self', *Journal of Consumer Research*, 15 (September), 139–168.
Bourdieu, Pierre (1984), *Distinction: A social critique of the judgement of taste*. Cambridge, MA: Harvard University Press.
Duesenberry, James S. (1949), *Income, Saving and the Theory of Consumer Behaviour*. Cambridge, MA: Harvard University Press.
Ekström, Karin M. (2007), 'Parental consumer learning or keeping up with the children', *Journal of Consumer Behaviour*, 6 (July–August), 203–217.
Ekström, Karin M. (2013a), 'Om behovet av konsumtionskritik i ett konsumtionssamhälle', [The need for critique of consumption in a consumer society], in Lennart Weibull, Henrik Oscarsson and Annika Bergström (eds), *Vägskäl, 43 kapitel om politik, medier och samhälle, SOM-undersökningen 2012* [*Crossroads, 43 Chapters About Politics, Media and Society, The SOM-survey 2012*]. SOM-Institute, University of Gothenburg: SOM-report 59, 369–385.
Ekström, Karin M. (2013b), 'The discovery of relations to artefacts in the boundless process of moving', in Barbara Czarniawska and Orvar Löfgren (eds), *Coping with Excess: How organizations, communities and individuals manage overflows*. Cheltenham, UK: Edward Elgar Publishing, 173–191.
Ekström, Karin M. (forthcoming), 'Conformity and distinction in Scandinavia's largest department store', in Alejandro N. García Martínez (ed.), *Being Human in a Consumer Society*. Farnham, UK: Ashgate.
Ekström, Karin M. and Helene Brembeck (eds) (2004), *Elusive Consumption*. Oxford: Berg Publishers.
Ekström, Karin M. and Torbjörn Hjort (2009), 'Hidden consumers in marketing, the neglect of consumers with scarce resources in affluent societies', *Journal of Marketing Management*, 25(7–8), 697–712.
Ekström, Karin M. and Kay Glans (2011), 'Introduction', in Karin M. Ekström and Kay Glans (eds), *Beyond the Consumption Bubble*. New York: Routledge, 3–14.
Ekström, Karin. M. and Nicklas Salomonson (2014), 'Reuse and recycling of clothing and textiles – a network approach', *Journal of Macromarketing*, 34(3), 383–399.

The European Parliament and the Council of the European Union (2008/98/EC), 'Directive 2008/98/EC of the European Parliament and the Council on Waste and Repealing Certain Documents', *Official Journal of the European Union* L312/3, 0003-30.

Evans, David, Hugh Campbell and Anne Murcott (2013), *Waste Matters: New perspectives on food and society*. Oxford: John Wiley & Sons.

Gregson, Nicky (2011), *Living with Things: Ridding, accommodation and dwelling*, Wantage, UK: Sean Kingston Publishing.

Gregson, Nicky and Louise Crewe (2003), *Second-Hand Cultures*. Oxford: Berg.

Guagnano, Gregory A., Paul C. Stern and Thomas Dietz (1995), 'Influences on attitude-behavior relationships a natural experiment with curbside recycling', *Environment and Behavior*, 27(5), 699–718.

Hawkins, Gay (2006), *The Ethics of Waste: How we relate to rubbish*. Oxford: Rowman and Littlefield Publishers Inc.

Hjort, Torbjörn (2004) *Nödvändighetens pris – om knapphet och konsumtion hos barnfamiljer* [The price of necessity – about scarcity and consumption among families with children], Lund dissertations in Social Work 20. Lund: Lund University Press.

Hård af Segerstad, Ulf (1957), *Tingen och vi* [The things and us]. Stockholm: Nordisk Rotogravyr.

Jackson, Tim (2009), *Prosperity without Growth: Economics for a finite planet*. London: Earthscan.

Miller, Daniel (1997), 'Consumption and its consequences', in Hugh Mackay (ed.), *Consumption and Everyday Life*. London: Sage Publications.

Miller, Daniel (1998), *Material Cultures: Why some things matter*. London: The University of Chicago Press.

Wilk, Richard (2004), 'Morals and metaphors: The meaning of consumption', in Karin M. Ekström and Helene Brembeck (eds), *Elusive Consumption*. Oxford: Berg Publishers, 11–26.

World Commission on Environment and Development (1987), *Our Common Future*. Oxford: Oxford University Press.

PART I
Consumption and waste

1
RECYCLING THE HOME

The constant flow of domestic stuff, emotions and routines

Orvar Löfgren

Dreams of a perfect home

When you got up this morning did you notice the happy people gliding past outside – all dressed in the same practical white overalls and in a relaxed holiday mood, because they don't have to work more than three days a week now? There are high-rise apartments where helicopters are busy moving transportable apartment units to new locations when the family gets tired of the neighbourhood.

Knock on the door of a typical contemporary home. Look at the minimalist and practical interiors: there is no furniture, but beds and table and chairs emerge from the floor at the touch of a button. Go into the kitchen and see the electronic systems that have made cooking superfluous. Just press the button and your favourite dish is ready . . . This scenario is mainly taken from the book *How will we live in 2010*, written in the early 1970s (Hoyle 1972). Over the last years I have been interested in this genre of utopias of future domestic life and the forecasts for the new millennium that have been made since the 1950s. Many of these predictions worked out reasonably well for technologies of communication, but went totally wrong when it came to changes in habits, cultural values and emotions in domestic life (see, for example, Ashley 2011, Benford 2010 and Löfgren 2012). What is striking in much of this future gazing is the dreams of a minimalist home life, uncluttered, spotless and oh so practical. No friction. People themselves are unstressed, sophisticated, restrained – very cool. Over many decades there have been constant dreams of *the smart home*, so popular among future gazers, product developers and high-tech designers.

Now it is 2014 and the future is here. And it looks rather different. Let's take a glance inside a Swedish kitchen early one morning:

> The dried out plants on the windowsill, dead for months, slipping into the room in an unobtrusive way that made nobody think about throwing them

away. The table with its glasses and plates, the water jug with the water full of small bubbles of air; the dried crumbs lying around the places where the children have been sitting; the empty bags that the fruit arrived in, now resting like small hangars of plastic in-between stacks of drawings and drawing pads, felt and colour pens; and not to mention the two shelves on the wall next to the window, which were swelling coral reef-like over all the small things the kids had collected over the last years, from sweet dispensers formed like princesses or different Disney-characters, boxes with pearls, pearl boards, glue pens, toy cars, and water colours, to jigsaw pieces, Playmobile parts, letters and bills, dolls and some glass bubbles with dolphins inside which Vanja wanted to have when we were in Venice last summer.[1]

(Knausgård 2012: 260)

This is the author Karl Ove Knausgård describing his overflowing home in Malmö, Sweden, in the last of his six volume autobiographical novel, *My Struggle*. He reflects over the constant battle between chaos and order that goes on in Western homes and the ways that the material world is always about to take over.

The overcrowded home

It is this ongoing battle that I will explore here, part of a current project on tensions in contemporary domestic life being carried out with Billy Ehn. One tension has to do with different forms of overcrowding and the ways in which they interact or not. This chapter centres around three slippery verbs, *reduce, reuse* and *recycle* – slippery in the way that they are intensely cultural, and also in the ways in which they change meanings and directions in different contexts. They are also married to their counterparts in interesting ways. Reducing may also mean increasing (or vice versa), reusing may mean obliterating or discarding, while recycling is a process that may go in very different (and sometimes surprising) directions. The road towards sustainable living is full of paradoxes and sometimes counter-productive paths.

While many of the other chapters in the book focus on recycling, waste and waste management, I will look mainly at the earlier stages of the journey towards waste in domestic settings, the movements and transformations of stuff and the strategies for handling it. (The latter issue has been central in another project, *Managing overflow*, see Czarniawska and Löfgren 2012 and 2014.)

I will look here at three aspects of domestic life. First of all: *the stuff*. Although consumer studies have been influenced by both the material and the affective turns that have swept over social and cultural research during the last decades, it seems to me that there is still too little blood, sweat and tears in the ethnographies of domestic lives. Homes may be overflowing with semiotic signs, symbolic messages, dreams and longings, but they are, above all, full of objects, objects that constantly need to be handled.

Domestic life in the twenty-first century was supposed to be cyber-light and friction-free, thanks to all the new technologies that would simplify people's lives. Most Western homes are, however, still veritable jungles of clumsy objects and gadgets, utensils and tools crammed into every available space. Cupboards and wardrobes may be bursting, cellars and attics cluttered. Little gadgets let out green or angry red blips in the kitchen, electric cords create jungles under the tables. People devote a large amount of energy and resources to handling this abundance. Things are shuffled back and forth, rearranged, recycled. Every day, new objects enter Western homes and old ones are lost, forgotten or leave by the back door.

The second aspect has to do with the *habits and routines* of organizing and coping with domestic life that make things, activities and interests work together. 'What defines a home?' the anthropologist Mary Douglas (1991) once asked. Her answer is, not just a building with four walls, but also an internal order with rules, rhythms and morals. The home is a web of routines, silent agreements and ingrained reflexes about 'the way we do things here'. Words like 'routines' and 'habits' capture this in different ways. *Routine* is French for small road, paths carved in everyday home life, and habits make a home a habitat or at least habitable (see the discussion in Ehn and Löfgren 2010: 83–93 and Winther 2009).

Homes can be seen as laboratories for developing new routines as well as safe havens for clinging to old ones. There are techniques of synchronizing, multitasking and sequencing, as well as constant whole- or half-hearted attempts to gain control and install some kind of order.

The third aspect is about *affects and emotions*. The home is not only crammed with stuff, it is also overflowing with feelings. Passion, boredom, guilt, longing, nagging irritation, explosions of home rage, moments of bliss – all try to co-exist with and also charge material objects (like that ugly sideboard we inherited from your father) as well as the normal everyday activities. (Who turned down the thermostat again? Where is my cell phone charger? And what are these towels doing on the bathroom floor!)

It is the constant cohabitation, clashes and recycling of these three aspects that I will explore. In many ways the home is a good example of what Doreen Massey (2005) has called the throwntogetherness of everyday life. Affects, activities and materialities work together, reinforcing or transforming each other. Such recyclings can take many forms: the physical movement of transporting stuff elsewhere into new contexts – be it the short journey on to the next shelf or down into the basement. In a similar manner, it affects the search for material outlets and points of anchoring. In such domestic entanglements, stuff, feelings and routines are transformed into new uses or functions. Just think about the ways in which waste, junk and dirt are, in cultural terms, produced by processes such as displacement, sorting out and recategorization, as in Mary Douglas's classic credo: 'Dirt is matter out of place' (see the discussions in Douglas 1966 and Thompson 1979).

My material is a bricolage, based on ongoing fieldwork, interviews and observations, as well as a wide range of sources, from academic research to popular culture and fiction, as well as several surveys of contemporary homemaking.[2]

Stuff on the move

Looking around my own home, I am struck by how many of my possessions I fail to note, although some of them are right in front of me. They have been transformed into the driftwood of the consumer society, left stranded on the drawer, the top shelves in the kitchen or in wardrobe corners. How did this happen?

Someone puts a white ceramic bowl on the sideboard as a nice design accent. There it is, simple, beautiful and above all seductively empty. All of a sudden there is an empty matchbox in it, next to a couple of coins. The ice is broken and through a magic power new objects are attracted. A pair of sunglasses, an old lottery ticket, an unpaid electricity bill and some used batteries. Step by step a mountain is growing on top of the sideboard, until one day someone gives the living room a searching look: 'We can't have all this mess!' All bowls, tables and windowsills are de-cluttered. But, the objects silently bide their time; as soon as the back is turned they will take over again.

A tour of the basement and attic reminds me of all the half-finished projects stranded in there. In the kitchen cupboards I find relics of earlier campaigns of domestic moral rearmament: interesting spices and torn-out recipes that have never been tried, healthy ingredients never used and strange kitchen utensils which never worked. In a sense, all these ambitions that have been put on hold may keep a check on consumption. As unrealized daydreams or plans, they take up less space and costs than if turned into another resource-demanding project.

In domestic settings, things acquire a seductive peacefulness, as if they were just innocent stationary objects, but, as Jojada Verrips (1994) points out, they tend to have a life of their own. They decide to break down, move out of sight, hide or get lost. This mass of domestic objects creates friction, takes up space, gets in the way and is constantly on the move. How are these mental and physical micro-movements organized into trajectories and stations?

Knausgård continues to reflect over the mysterious ways in which stuff takes over:

> The shelf was a station; when the objects landed there, they were out of circulation, and stayed put. We had several similar stations where the life of things suddenly ended, especially the long bench in the passage ... both under and over the bench there were cupboards where all kinds of stuff and rubbish were lying next to things we needed, but no longer remembered we had.
>
> (Knausgård 2012: 261)

He continues to list all these stacks of stuff gathering in the nooks and crannies of the apartment. His description can be compared to the anthropological study of thirty-two Californian homes, *Life at Home in the Twenty-First Century*, in which a team of researchers make detailed ethnographies of domestic life and domestic stuff (Arnold et al. 2012). The first household assemblage they analysed had 2,260 visible possessions in the first three rooms that were documented (two bedrooms and the living room), not counting all the stuff out of sight in lockers, closets and drawers.

The people interviewed often complained about their homes 'being a mess'. Just as in Knausgård's home, there were stations in which stuff piled up, or 'dumping grounds' as someone called them. Storage spaces developed everywhere, often quite unplanned, like the garage, where there was no longer room for a car, the bedroom corners and especially unused in-between spaces. Other home surveys show other storage inventions, such as the unused bidet that had become home to a crowd of shampoo bottles and cosmetics.

Just like the white bowl, tables are inviting empty spaces waiting to be filled up. In the Californian study, kitchen tables emerged as surfaces on which stuff flowed in and out during the day, reflecting the fact that this was the most used domestic space of all, home for a wide range of activities.

Domestic stuff often is described in terms of overflowing fluids: stuff 'spills out', creates 'waves' and 'driftwood', or builds up 'coral reefs'. A uniting theme in many of these surveys is that toys have emerged as the strongest kind of driftwood. Of all the domestic overflows talked about in families with small children, toys often stand out (see, for example, Brembeck 2014 and Plowman and Stevenson 2013). A new child in an American family may result in a 30 per cent increase of stuff at home (Arnold et al. 2012). 'Stuff – more than anything else meant toys', writes Nikolaj Zeuthen in his ethnographic novel about a young Danish family. Like Knausgård, he goes on to list the children's belongings, from dispersed Lego sets and Barbie dolls to toy animals and plastic trumpets. Once they were distinct objects, bought on special occasions, but now, he writes, they had turned into a shapeless mass, 'flowing like the waves against the beach, moving in and out of cabinets, in and out of boxes and baskets, down and up again on the walls, always with Rune and Anna as those who had to restore order, stooping with stiff backs' (Zeuthen 2012: 113).

In families with small kids there are complaints of toys drifting all over the place, helped by the fact that the children do not want to play in their room but drag all kinds of stuff into every space, wanting to be where the action is. The whole home is turned into a playpen, leaving the play-room unused.

Toys and other stuff adrift also highlight another form of recycling: the coming and going of objects. Things disappear easily; they move out of sight, hide in the clutter of Knausgård's cupboard or get lost in the garage. What kind of recycling is found in the processes of loss, disappearance and retrieval?

Moving out of sight

An exhibition of lost and found things, 'This could be yours' in Copenhagen 2006, illustrates this by using lost objects that had been gathering in storage at the Copenhagen Transport Authority. In the catalogue, the Danish poet Mårten Søndergaard addresses the mysteries of disappearing objects:

> The things. I can't find them. They are gone. They could be there or there … But they aren't. They are somewhere else. The world has moved four millimetres

and a crack has emerged. The crack is widening and more stuff falls through it. Now they are in a different place. Unreachable, invisible and bordering on the insane. As I find them again, at the last moment, I wonder: where have they been? Maybe they have been to God for an overhaul. But now that they are back, are they blessed or cursed? I don't know. I have to take care. When they return through the four millimetre crack they look shinier than before. With a smell of lemon.

In the introduction, the curators point to the different emotional reactions to losing belongings. Why is it that some disappearing objects create acute feelings of loss, while others just go into hiding, unnoticed, like the many forgotten objects in Knausgård's cupboard? There is the woman whose husband routinely reorganized the home video collection, discarding some of it but accidently erasing the film of their daughter's baptism. His wife could not forgive him:

> I still wake up at night and cry. I never got to show my daughter her name-giving ceremony. It feels like I have lost a part of life and I don't really know how to get over this loss. Maybe it sounds pathetic to cry over a video recording, but it meant more than that.

Losing stuff in the home can be emotionalized in several ways: the daily frustration of searching for necessities, the keys, the salt, the missing shoe, the favourite toy. Search and find missions have their own temporalities and spatialities: for families with children, the morning exodus is a trying time. Then there are the things that get lost without being noticed, or even with a sigh of relief. The object in some mysterious way got rid of itself. Instead of actively discarding stuff, people often just let go … In the same way, emotional reactions vary when things are found again: ranging from acute happiness to total indifference – 'who needs this old thing, let's throw it away properly this time!' (See the discussion in Ehn and Löfgren 2007: 190–196.)

Kevin Hetherington (2004) has problematized the trajectories (mental and physical) of possessions and points out that the classic chain of *production – consumption – disposal* does not follow in an inevitable, discrete, linear temporal sequence (Hetherington 2004: 168). How are objects transformed, by moving back and forth, changing both position and value? He compares objects with the institution of double burials found in some cultures, a two-stage ceremony conducted to help the bereaved adjust to the new situation. There are the 'first burial places' of a discarded object – the refrigerator, the basement or the digital 'wastepaper basket' of the computer ('Do you really want to delete this?') – stations that give people a chance to adapt to the final discarding in trash cans or in cyberspace.

An unfinished disposal may create a situation of haunting, a vague feeling of debt, guilt or loss, as Hetherington points out. Objects not only haunt us, they may also surface and return in different ways. As Judith Attfield (2000) puts it in her book *Wild Things*, some materials are better for this than others. Her main example is personal clothing, objects that can carry memories and emotions in strong ways. The same goes for media memorabilia (see Löfgren 2012: 120–124).

Lost objects may thus continue to clutter up both people's homes and minds. Disappearances may take the form of accidents, negligence or oblivion, existing in a tension between the active or passive. Objects retrieved may return with a new aura or new functions. In situations like the breaking up of homes, as in separations and estate inventories or just in moving house (Ekström 2013), rediscoveries and reappraisals are made. There are complex processes of recycling at work here.

Emotions on the move

> *Amathophobie* – fear of dust, *Ataxophobie* – fear of disorder, *Atelophobie* – fear of the unfinished, *Kainophobi* – fear of the new, *Kenophobie* – fear of emptiness, *Pentheraphobie* – fear of mother in law …
>
> *(Domestic phobias listed in Das Buch der Ängste, Schmidbauer 2007)*

The question of loss and retrieval actualizes the question of emotional flows and transformations in the home. In the eighteenth century, people fantasized about emotional landscapes; they imagined fictional worlds such as the sea of boredom, the island of happiness, the dark woods of despair, the road of hope (see Bruno 2002: 205–245). Maps like those can be drawn of contemporary homes, instead of just furnishing plans. Where, why and how do we find the flows of emotions and changing moods in an apartment or a house – and how do they change with the rhythms of day and night, workdays and weekends? Emotions are usually on the move, changing shapes and directions, and finding moorings, hiding places or temporary refuge.

Check the atmosphere or mood of the living room at night or in the kitchen in the morning. Where do irritations gather? What are the spaces for daydreaming or blissful relaxation, moments of happy togetherness or a creeping feeling of boredom and frustration? Emotions may be stored in kitchen cupboards or in a piece of furniture, harbouring old resentments or blissful memories. Different moods change the interior and the furniture. Melancholia wraps the whole home into a grey mist, instead of the rosy light of blissful moments. In a novel by Jenny Offill, the wife finds out that her husband has another woman. She feels queasy and retreats into the bathroom:

> The longer she sits there, the more she notices how dingy and dirty the bathroom is. There is a tangle of hair on the side of the sink, some kind of creeping mildew on the shower curtain. The towels are no longer white and are fraying at the edges. Her underwear too is dingy, nearly gray. The elastic is coming out a little. Who would wear such a thing? What kind of repulsive creature?
>
> *(Offill 2014: 115)*

The bathroom is ready to amplify her mood of decay and depression. Homes can change rapidly from inviting and warm to drab and unfriendly. Stress may make the kitchen seem hostile. In her 2004 novel *Ta itu* ("Take apart"), for example, Kristina Sandberg describes a young mother's nervous breakdown. The main character finds that she can't cope with all the demands and expectations that both she and those around her are posing. Sometimes it seems as though she is being aggressively scrutinized by everything around her. Even the dust and fluff whirl accusations into the air:

> Pack, clean, make the dinner, take care of the plants, wash those dirty windows highlighted by spring's merciless sunshine. Anders will be late. Let's hope the children will behave themselves. I must clean out the fridge, then there's dinner, fish fingers and mashed potatoes.
>
> *(Sandberg 2004: 30)*

Everything gangs up on her. As soon as she lights a cigarette to calm her nerves, her son accusingly waves a brochure about stopping smoking. Her mother-in-law calls with unwanted advice about cleaning. The homemade marmalade cake decides to sink in the middle and the icing turns into a puddle; the fridge door is all sticky, crumbs spread themselves all over the place and the kitchen smells of burning fat.

In his study of this struggle people have with objects, Jojada Verrips (1994) argues for the emergence of modern forms of animism: 'The damn thing didn't do what I wanted it to do!' Objects bought to make lives easier also make life more complicated. They put people to the test when they decide to give us trouble and stop working or go into hiding somewhere. People are driven to the verge of fury or tears at one time or another when they fail to reprogramme the DVD recorder, when the computer screen freezes or when the washing machine turns whites to coloureds. Gadgets are handled brusquely, furniture kicked or kitchen utensils thrown on the floor or at other family members.

Objects and affects come together in many ways. Why is it that some things attract certain feelings and become a focus of irritation, happiness or sadness? Or, alternatively, how do affects cling to certain objects? In a discussion of 'happy objects', Sara Ahmed (2010) explores such processes of stickiness, whereas Sianne Ngai (2005) explores how irritation is materialized. Kathleen Stewart (2010) has explored the 'mood work', in the entanglements of objects, spaces, mind-sets and movements.

Never good enough

An important domestic feeling is guilt at not having a good enough home or family life, with control and order lacking. In the interviews with Californian families, the theme of messiness occurs frequently, mainly among the wives:

This is the office. It's a total mess. We probably should, you know, organize it better ... And here we have the garage, with everything. It is usually a total mess and it's a total mess today again. This is where we have bikes and all the old furniture, sofas and things we don't use. It is, how can I say it, it's a mess. It's not fun, it should be cleaned up and we should probably get rid of a whole bunch of stuff.

Nicolaus Zeuthen captures the strong emotional charges of the attempts to organize everyday life:

> They never felt at peace. The home was never orderly enough and demanded constant tidying and cleaning and shopping trips where the crazy idea was that new toys and new baskets, boxes and cabinets maybe could create better order and make life easier ...
>
> In the apartment everything felt unsettled and undifferentiated. The classic Wegner rocking chair had never found its proper place, but was only in the way of the bookcase, where the books stood unlovingly without order. The small Klint sofa table was painted over by the children's felt pens and stood ashamed against a wall and was like most other tables in the home - no more a piece of furniture around which one gathered and concentrated on something, but just a surface for storing stuff.
>
> *(Zeuthen 2012: 112)*

Knausgård has a similar take. The stuff piling up in the apartment could give his wife panic-like attacks:

> it was the feeling of chaos it gave her, which she couldn't handle. Often she came home with storage utensils, which should sort of organize everything; different boxes for different things, a tray for my post, one for hers, marked with our names, as she had seen at other people's places who seemed to be orderly, but the systems collapsed after a few days, and everything flowed out again as before.
>
> *(Knausgård 2012: 262)*

Knausgård also embarked on projects of de-cluttering, but had to give up. It was as if the things 'were alive, as if they lay there and pulled stuff towards them in order to grow and be powerful'. He keeps reassuring himself that this was not a moral issue:

> We were not bad people, even if we were messy. It was not a sign of bad morals. This I tried to say to myself, but it didn't help, the feelings were too strong; when I walked around in the mess, it was as if it accused me, accused us, we were bad parents and bad people.
>
> *(Knausgård 2012: 262)*

A theme running through many of the battles of overflowing stuff in the examples I have used is a nagging feeling of being stuck with too much of it. There is the constant dream of a simpler, or even a minimalist home, and there are many (often half-hearted) attempts at reform, at consuming less and getting rid of more and becoming a better-organized household. There is the constant barrage of images of good or beautiful living in home styling magazines and IKEA catalogues, or fantasies about the perfect homes of neighbours. Questions of guilt and the gap between ideals and reality are closely tied to the constant visits of invisible guests, those imaginary judges or censors that tell people what a perfect or good home should look like. In an increasingly complex world of cohabitation arrangements, the ideal of the nuclear family still stands strong. Guilt is thus a good example of the agency of emotions, often on an unconscious level. Guilt may transform the home, present it in certain lights, demanding certain activities or blocking others.

To changing ideals of trendiness, beauty, harmony, control and tidiness are now added the ambitions for more sustainable living. There is no lack of advice for green living on the internet, in the press and in a vast quantity of self-help literature. As Richard Wilk points out in the Afterword, much of this well-meaning advice does not work and instils more of a vague feeling of guilt rather than actual changes in everyday habits. He points out that the moralizing tone can become counterproductive. A recent bestseller in the genre illustrates this dilemma. I am thinking of Bea Johnson's *Zero Waste Home: The Ultimate Guide to Simplifying Your Life by Reducing Your Waste* from 2013. (There will be many more such 'ultimate guides' to come.)

Reading her torrent of advice can leave anyone a bit exhausted. The aim is good, but so many of the hints and suggestions seem time-consuming, complicating domestic life rather than simplifying it. Sometimes the book reads like the old genre of thriftiness guides for housewives from the 1910s or 1950s: how to make your own notebook or soap, and advice on saving free samples of everything. It is a good example of how the ambition to reduce results in 'less becoming more'. We need to continue to explore which campaigns for sustainable living are effective and what arguments people will listen to.

Work at home – an entanglement of tasks, gadgets and emotions

The worries about clutter and overconsumption illustrate a general trend. Over the last decades, homes have become more open and boundaries between activities and rooms more fluid. This is not only the result of open-space planning and doing away with doors and walls, or opening up the kitchen to other areas. In older homes, activities and people also mingle in new patterns (which also results in a new longing for privacy and a yearning to close the door behind you).

There are many tensions like these building up in today's homes and the most striking is perhaps the mixing of work and leisure. In laptop families all over the world, office work has invaded the home in rapidly changing ways. Work, leisure and parenting are mixed; people write memos at the kitchen table surrounded by the kids, or answer emails in front of the TV (see, for example, Darrah et al. 2007).

The digital home emerged slowly in the 1980s. Gradually, in the 1990s, an electronic jungle began to take over. Ugly black and grey plastic invaded home interiors in successive waves: camcorders, PlayStations, laptops, pocket games, cellphones, MP3 players, transformers, adapters and batteries. New rhythms of media use and multi-tasking emerged. (In 2013, for example, Americans checked their smartphone an average of 150 times a day; Cheshire 2013.)

By 2014, work had returned to the home in many settings, helped by mobile media. On one and the same family sofa, dad can be surfing the internet, mother answering emails from work on her smartphone, while the kids are online gaming and the toddler is trying out the iPad. All kinds of improvised workspaces emerge as the job invades the home: laptop work goes on in the bedroom, at the kitchen table or even on the ironing board.

In her study, *Work's Intimacy*, Melissa Gregg (2011) explores the conflicts and discussions that the constantly moving boundaries of working at home can produce. When, where and how is it OK to work and for whom? 'Laptop at dinner, that's where I draw the line' or 'Why is it that I will organize my 100 latest emails on the sofa at home, but never at work?' 'The kids say we are hardly there, just hooked on to the screen.' This is a battlefield with strong emotional charges.

Wanda Orlikowski (2012) followed the introduction of BlackBerry® phones in an American company. They were introduced to facilitate and speed up emails and other forms of communication, but there were no formal rules issued about their use. She looked at how expectations of availability and answering slowly developed. How fast did you have to return an email? Should it be done after work, and if so when? People tried to set up their own rules, such as 'I never answer anything after work or at least not after 10 pm', but slowly work communication invaded domestic life and leisure. Some people even began answering messages in bed and their frustrated partners tried to find vacation spots that had no BlackBerry® coverage. By following changing practices over months, Orlikowski could study the social dynamics of a workplace culture changing, without there ever having been any master plan. What happened here was that work invaded new spaces and contexts in the home and the bed suddenly became a workspace at midnight, or the bathroom the place for important work calls. Where did the office end and the home start? Categories, tasks and spaces were blurred.

Domestic rhythms and cycles

One further way of approaching these flows of stuff, habits and emotions is to look at the lifecycles of homemaking. At what stages are homes recycled, reinvented or just redecorated? Let me give a few examples.

When a couple moves in together it is possible to follow the institutionalization of 'our routines', which also works as a set of rituals of togetherness, but there is also the battle about 'your bad habits and my good ones'. The transition to a family with small children calls for all kinds of renegotiation of activities, space and priorities. When children leave home, there is the empty nest syndrome, not only the emptiness of the empty spaces but also of routines that no longer make sense. Divorce recycles the home in many ways, just as remarriage brings coordination problems with visiting kids and temporary sleeping arrangements, a bigger car, more stuff.

In old age, the home shrinks, most of the stuff surrounding people becomes a backdrop, sofas that are never used, books not read and china never put on the table. In the end, there might be an island of a TV chair and a remote control, a small table for frozen dinners and next to that the bed – the rest of the home is hardly used.

Exploring the lifecycles of homes and families highlights changes that expand or accelerate consumption or diminish it, as the home takes on new functions. We need to look at how different rhythms, practices and lifecycles co-exist, clash with or block each other.

The varying temporalities and trajectories of the lifecycles of stuff and routines create other synchronization problems. When and how do domestic objects and habits start looking tired or unfashionable? – a question Colin Campbell addresses in his chapter on the cult of newness. Questions of waste are, of course, often more about cultural wear and tear rather than just physical wearing out. There are processes of fashion cycles and commodification at work here, some of which I have discussed elsewhere. In a historical perspective, we can see how the fashion logic developed step by step in Parisian *haute couture* was imported into the marketing of domestic stuff (see the discussion in Löfgren 2005), changing the cultural lifespan of objects. Gradually, the home turned into *L'Empire de l'ephémère*, to paraphrase Gilles Lipovetsky (1987), with accelerating fashion cycles colonizing new territories and the cultural age span of domestic stuff shrinking. What is the life span of a kitchen (IKEA knows), a Barbie doll (Mattel knows), or a hammer, a cardigan, a set of china? What are the processes that lead up to the feeling of a setting or an object suddenly looking tired or unfashionable? Look at this kitchen (or sofa or wallpaper), we will have to do something about this!

Habits also have their lifecycles, starting with activities that gradually become cemented into routines (Campbell 1996). As new commodities and technologies are introduced into the home, they may clash with or have to accommodate to ingrained habits.

There is also the synchronization of the gap between domestic ideals and everyday realities. Most people live in apartments or houses designed for older ideals, like the functionalist emphasis on small kitchens (where the housewife was the only one doing the cooking), the grand master bedroom or the Californian dreams from the 1960s of a garden activity area where now the barbecue sets and jacuzzis stand unused. The longings for the perfect home or the ideal family can be materialized in sudden projects of home improvement or buying sprees, signalling a family on the move, pulling members together in a shared project. (The brand new kitchen should be the scene of more family dinners and home-cooked meals, not the usual take-away stuff eaten on the run.)

As objects, activities and moods move around inside the house, erasing traditional boundaries or challenging established categories, they are redefined, just as the spaces and situations they enter into are as well.

Take the ironing board. In the 1970s, women found that ironing worked well with watching the TV. The ironing board thus entered the living room. Today the ironing board can be used as an improvised office desk for laptop work, just like beds have become multi-media stations for cellphones and iPads, books and snacks, which partly accounts for the fact that they are getting bigger and bigger – they have to handle more stuff. The fridge door has turned into a communication centre, for appointments, to-do lists and short messages.

A domestic moral economy

The concept of the home as a moral economy is one way of integrating many of the processes I have explored. Returning to Mary Douglas's classic paper, she discusses the home as an entanglement of conventions and totally incommensurable rights and duties. What she describes is very much a moral economy, constantly tackling questions of solidarity, sharing and assistance, as well as questions of fairness. The home has to synchronize not only tasks and activities but also needs and longings. (This goes not only for family homes but single households as well; it is about all that makes a home different from a lodging or a hotel.)

It is a moral economy that produces many tensions, for example between individual aspirations and activities and 'the family or household good'. There is a diffuse 'we' often hovering in the background. 'Do "we" really need a new TV, a bigger house, dessert for dinner?' Home is a site of negotiation, with constant wheeling and dealing, trying to make different priorities and interests co-habit. The author Jenny Diski describes breaking up a relationship and reclaiming her home:

> It is almost as a dance, a floating self that breathes its way around the place while you only seem to brush your teeth and make cups of tea. It is a celebration of solitude – but also of control, no need to synchronize.
> *(Diski 1999: 213)*

Life at home is often seen in opposition to public life, for example as a moral battle about how far market commodification should be allowed to expand. Recent years have witnessed a heated debate about the commodification and outsourcing of domestic activities in middle-class homes. Is it OK to invite into the home the services of a nanny, a children's birthday organizer, a family coach, an interior decorator or a stylist? (See the discussion in Hochschild 2011.) Such discussions can tell us about contested boundaries and norms of what a home should be, or what it should do on its own – without market interference and hired help.

It is, however, important to note that there is no unilinear steamrolling process of commodification. In domestic life we can also follow the decommodification processes of objects and activities, which are withdrawn from a market logic (see Löfgren 2013). Think, for example, of new trends in sharing, swapping and borrowing, which are discussed in this volume by Belk and others.

The moral economy of the home also reflects different positions, and thus engages questions of class, gender and generation. In some ways, the role of the home as a moral economy is becoming an increasingly important issue. There are more negotiations of what is expected of household members, of 'what is fair or not', which is linked to the processes of increasing individualization in modern homes, with a greater emphasis on 'my room, my taste, my priorities and my privacy' among both children and adults.

The moral economy of a given home is rarely visible in grand declarations about rules, rights and duties. It is hidden in mundane situations, which explains why seemingly trivial routines or actions can all of a sudden result in a flare of emotions. Power and hierarchies are reinforced or challenged. Children in divorced families moving between Dad's and Mum's new homes learn about small but important shifts in moral economies (see Winther et al. 2014).

A dense and diffuse moral economy will always risk being overflowed by the constant clashes between stuff, routines and emotions. Questions of sustainability add another powerful moral dimension.

At the recycling station

The throwntogetherness of stuff, practices and feelings inside the home has been the focus here. By looking at such entanglements, we might learn something about waste and its generation, as well as about sustainability. I am reminded of this when I visit my local recycling station with a carload full of what once were cherished belongings, remnants of half-finished home improvement projects, once trendy stuff now out of fashion. Inside the black plastic bags, they are all now transformed into waste. Loading the car and going through the attic and the corners of cabinets, there are mixed feelings at work. Should I keep this? Maybe I should give it to someone?

At the station I am confronted by new categories. Should this old chair go into the container for 'wood – not impregnated', 'upholstered furniture', the charity shed or just end up in 'burnable'? Once inside the containers, the objects look embarrassed, abandoned and unloved. Watching other recyclers, I can sense different moods. There might be some guilt at throwing out some perfectly fine stuff, showing oneself up as a sloppy over-consumer. There is the relief of finally letting go, a load off your consciousness as well as your shoulders as the bags land with a bang in the container. Some people hesitate, 'maybe I could give this to the charity that has its own shed at the recycling station', but the shelves over there are already bursting with stuff. There is also a sense of moral victory hovering in the air – we may be over-consumers but we are good recyclers! Lynn Åkesson (2012) and Kathleen Stewart (2010) have explored this moral terrain in other contexts.

At recycling stations all over the world, everything is being recycled, emotions, routines, stuff and morals as well as ideas of homes. To understand the processes leading up to this moment we have to continuously watch the ongoing flow at home. Accommodating all the activities and projects, dreams and disappointments calls for constant domestic creativity, but also produces many half-hearted attempts and unfinished projects. This is nothing new, modern history is full of such tensions, and in the background there are the science fiction dreams of a perfectly minimalist and rational way of living, of a home under full control.

Notes

1 My translation from the Norwegian original.
2 One of these surveys is a recent extensive comparison of life at home in six nations made by a home furnishing company, which I draw on heavily.

References

Ahmed, S. (2010) 'Happy objects', in M. Gregg and G. J. Seigworth (eds), *The Affect Theory Reader*, pp. 29–51. Durham, NC: Duke University Press.
Åkesson, L. (2012) 'Waste in overflow', in B. Czarniawska and O. Löfgren (eds), *Managing Overflow in Affluent Societies*, pp. 141–154. New York: Routledge.
Arnold, J., Graesch, P. A., Ragazzini, E. and Ochs, E. (2012) *Life at Home in the Twenty-First Century: 32 Families open their doors*. Los Angeles: The Cotzen Institute of Archaeology Press.
Ashley, M. (2011) *Out of the World: Science fiction but not as you know it*. London: British Library.
Attfield, J. (2000) *Wild Things: The material culture of everyday life*. Oxford: Berg.
Benford, G. (2010) *The Wonderful Future That Never Was*. New York: Hearst Press.
Brembeck, H. (2013) 'Managing inflows, throughflows and outflows: Mothers navigating the babystuffscape', in B. Czarniawska and O. Löfgren (eds), *Coping with Excess: How organizations, communities and individuals manage overflow*, pp. 192–215. Cheltenham: Edward Elgar.
Bruno, G. (2002) *Atlas of Emotion: Journeys into art, architecture and film*. New York: Verso.
Campbell, C. (1996) 'Detraditionalization, character and the limits to agency', in P. Heelas, S. Lash and P. Morris (eds), *Detraditionalization: Critical reflections on authority and identity*, pp. 13–27. London: Blackwell.
Cheshire, T. (2013) 'Now hyperstimulation is making you smarter', *Wired*, Dec 2013: 110–117.
Czarniawska, B. and Löfgren, O. (eds) (2012) *Managing Overflow in Affluent Societies*. New York: Routledge.
Czarniawska, B. and Löfgren, O. (eds) (2014) *Coping with Excess: How organizations, communities and individuals manage overflow*. Cheltenham: Edward Elgar.
Darrah, C. N., Freeman, J. M. and English-Lueck, J. A. (2007) *Busier than Ever: Why American families won't slow down*. Stanford, CA: Stanford University Press.
Diski, J. (1999) *Don't*. London: Granta.
Douglas, M. (1966) *Purity and Danger*. London: Penguin.
Douglas M. (1991) 'The idea of a home: A kind of space', *Social Research*, 58(1): 287–307.
Ehn, B. and Löfgren, O. (2007) *När ingenting särskilt händer*. Stehag: Brutus Östling.
Ehn, B. and Löfgren, O. (2010) *The Secret World of Doing Nothing*. Berkeley, CA: University of California Press.
Ekström, K. M. (2013) 'The discovery of relations to artefacts in the boundless process of moving', in B. Czarniawska and O. Löfgren (eds), *Coping with Excess: How organizations, communities and individuals manage overflows*, pp. 173–191. Cheltenham: Edward Elgar.
Gregg, M. (2011) *Work's Intimacy*. London: Polity.

Hetherington, K. (2004) 'Secondhandedness: Consumption disposal, and absentpresence', *Environment and Planning D: Society and Space,* 22: 157–173.
Hochschild, A. (2011) 'Emotional life on the market frontier', *Annual Review of Sociology,* 37: 21–33.
Hoyle, G. (1972) *2010: Living in the Future.* London: Heinemann.
Johnson, B. (2013) *Zero Waste Home: The ultimate guide to simplifying your life by reducing your waste.* London: Particular Books.
Knausgård, K. O. (2012) *Min kamp.* Vol. 6. Oslo: Forlaget Oktober.
Lipovetsky, G. (1987) *Le Bonheur paradoxal: Essai sur la société d'hyperconsommation.* Paris: Gallimard.
Löfgren, O. (2005) 'Catwalking and coolhunting: The production of newness', in O. Löfgren and R. Willim (eds), *Magic, Culture and the New Economy,* pp. 57–71. Oxford: Berg.
Löfgren, O. (2012) 'It's simply too much! Coping with domestic overflow', in B. Czarniawska and O. Löfgren (eds), *Managing Overflow in Affluent Societies,* pp. 101–124. New York: Routledge.
Löfgren, O. (2013) 'Changing emotional economies. The case of Sweden 1970–2010', *Culture and Organization,* 19(4): 283–296.
Massey, D. (2005) *For Space.* London: Sage.
Ngai, S. (2005) *Ugly Feelings.* Cambridge, MA: Harvard University Press.
Offill, J. (2014) *Dept. of Speculation.* New York: Knopff.
Orlikowski, W. J. (2012) 'Sociomaterial practices: Exploring technology at work', *Organization Studies,* 28(09): 1435–1448.
Plowman, L. and Stevenson, O. (2013) 'Exploring the quotidian in young children's lives at home', *Home Cultures,* 10(3): 329–348.
Sandberg, K. (2003) *Ta itu.* Stockholm: Nordstedt.
Schmidbauer, W. (2007) *Das Buch der Ängste.* München: Blumenbar.
Thompson, M. (1979) *Rubbish Theory: The creation and destruction of value.* Oxford: Oxford University Press.
Stewart, K. (2010) 'Afterword', in M. Gregg and G. J. Seigworth (eds), *The Affect Theory Reader,* pp. 213–238. Durham, NC: Duke University Press.
Verrips, J. (1994) 'The damn thing didn't "do" what I wanted: Some notes on modern animism in western societies', in J. Verrips (ed.) *Transactions: Essays in honor of Jeremy Boissevain,* pp. 35–52. Amsterdam: Het Spinhuis.
Winther, Ida Wentzel (2009) 'Homing oneself: home as a practice', *Haecceity Papers,* Vol. 4.
Winther, I. W., Paludan, C., Gulløy and M. Middelboe Rehder (2014) *Hvad er søskende?* Copenhagen: Akademisk Forlag.
Zeuthen, N. J. (2012) *Verdensmestre. En historie fra 00'erne.* Copenhagen: Samleren.

2
THE CURSE OF THE NEW

How the accelerating pursuit of the new is driving hyper-consumption

Colin Campbell

Introduction

Environmental philosophers and commentators have long argued that the nature of Western civilisation is such that significant cultural obstacles stand in the way of fostering a truly sustainable relationship with planet Earth. Indeed the Western worldview as a whole is frequently understood to be at the heart of contemporary environmental problems, mainly because of its 'materialistic', 'reductionist', 'disenchanted' and 'dualistic' character (Berry, 2006; Devall, 2001; Devall and Sessions, 1985; Moncrief, 1970; White, 1967). However, in this chapter I want to concentrate on one ingredient of modern Western culture that can be considered to constitute an equally serious obstacle to developing a truly sustainable relationship with the natural world, indeed one that has a good claim to be the major force behind the current phenomenon of hyper-consumption – that is to say consumption that appears to be undertaken as much for its own sake as to meet any real need – and yet one that is rarely given much prominence in these discussions. This is the high value attached to the new and the novel. While accepting that this feature of contemporary culture does not account for all the forces responsible for the extraordinary high levels of consumption that characterise modern developed societies as well, increasingly, of that of many rapidly developing ones, it is nonetheless suggested that it is responsible for the major part of this tendency, and thus, if indirectly, the unsustainable nature of the contemporary Western way of life. The various forms taken by this enthusiasm for the new and novel will be examined, together with evidence of the accelerating character of the consumption of products regarded as manifesting these qualities (Hartmut, 2013). Finally, some consideration will be given to whether there is any realistic possibility of curbing these tendencies, if not actually diminishing the significance attached to the new and novel in general.

Three forms of consuming the new

In an earlier paper, I outlined three different senses in which one might talk about something being new (Campbell, 1992). The first is the idea of the new as the fresh or newly created; the second the new as the improved or innovative; while the third is the new as the unfamiliar or novel. Now in using the word 'new' to mean fresh it is being opposed to 'old' in the sense of something that is worn or merely aged. This is the sense in which we talk about a new moon, a new baby or of new shoots on a bush or plant. In none of these contexts does the word 'new' imply anything novel or different from what went before. The new moon is in effect very familiar and so are babies and shoots. The contrast here is purely temporal, referring as it does to monthly, generational or seasonal change. The basic assumption is that all things age with time, hence requiring regeneration if they are to continue; and we use the word 'new' to indicate that regeneration has indeed taken place. The second sense of 'new' relates more to efficiency and technical capacity than newness in this purely temporal sense. Here the 'new' is the improved, the innovative or simply the most developed, in a long line of objects, processes or practices that have succeeded each other over a number of years. Commonly products that are presented as new in this respect embody the present state of the art in the relevant technology, and hence reflect the most recent scientific or technical knowledge or expertise. Finally, there is the third sense in which the word new is used, one that refers neither to the fresh or the improved but to that which is novel or unfamiliar. Here the contrast is purely experiential and therefore differs significantly from the two previous uses. For while the novel may also be new in the sense of the freshly created, this does not have to be the case, as objects that are old may still be unfamiliar to the person encountering them. Similarly, while some new and improved objects or experiences may also strike people as novel, many will strike individuals as fundamentally familiar.

Now a consideration of these three forms of the new leads logically to three different forms of consumption, ones that we can label replacement consumption, innovatory consumption (or rather consumption of products that are seen to be innovatory) and fashion or aesthetically-related consumption; distinctions, which, although analytically useful, frequently overlap in practice. For example, a consumer may come to replace an item that is broken, such as a TV or computer, to find that the only models available on the market are up-dated or improved versions of the original. Equally the most current or up-to-date model of a product may, for that very reason, also be the most fashionable. Despite this, it is suggested that these distinctions are analytically useful. Consequently, each of these three forms will now be considered in relation to their possible contribution to the contemporary phenomenon of over-consumption.[1]

Replacement-driven consumption of the new

Replacement consumption can be subdivided into two main types, at least in principle. First, there are those products that are 'disposable', which is to say designed to be used only once, and consequently require replacing immediately after use. Then, second, there is the replacement of objects that are designed to have a reasonably long life but become broken, damaged or faulty in some respects, and therefore, if deemed unrepairable, have to be replaced. There is also a possible third kind, one that stems from a change in the consumer's circumstances such that the product in question no longer fulfils the need that originally prompted its purchase, and hence a new product is needed to satisfy the same need. An obvious example of this latter form of replacement consumption would arise from the fact that children have a tendency to outgrow their clothes and therefore the need arises for them to be replaced by ones of a larger size. Another example would be the need, on the part of members of the older generation, for new glasses to replace those they currently use as their eyesight deteriorates with age. However, as these two examples would suggest, replacement consumption of this kind tends to conform to a well-established and traditional pattern and hence there would seem little to suggest that there might have been any significant change in its overall incidence or extent in recent years. Consequently, the following discussion will focus on the other two types of replacement consumption.[2]

Disposable replacement consumption

The first type of replacement consumption identified above relates to single-use or disposable products, that is to say products that are either used up completely soon after purchase or, if set aside for later, have all their utility exhausted completely in what is, in effect, a single act of consumption. Essentially this is replacement consumption in the sense of re-supply and hence one naturally tends to think, first of all, of foodstuffs, given that these are necessarily used up in the very process of consumption, and consequently most people are forced to engage in the replenishment of such goods on a regular basis in the form of the weekly, monthly or even daily, shop. But of course it is not only food and drink that come into this category as the number and range of products that come under the heading of 'disposable' is considerable. That is to say, the number of products that are designed, and indeed generally advertised, as single use, and which consequently effectively have no second-hand value.[3] Many of these will also feature in the regular shopping for groceries mentioned above, such as soaps and shampoos, dishwasher tablets or liquids, vacuum cleaner bags, candles, matches or medicines. Yet the range of products that fall into this category is far greater than might be suggested by an examination of the contents of the trolley of an average supermarket shopper. Indeed, one can now find disposable products for sale in almost all product areas. By way of illustration, the following is just a small sample of all those products that are currently advertised as either 'disposable' or 'single use': aprons, barbecues, cameras, dehumidifiers, ear plugs, flip flops, gloves, hair dye, ice packs, jelly moulds, knickers, lighters, mascara, nappies, overalls, ponchos, quiche dishes, razors, slippers, toilet seat covers, umbrellas, video cameras, wine glasses, x-ray markers, yoga mats and zip handcuffs.[4]

It is also worth noting that it is not just in the context of purchasing products that consumers may encounter the phenomenon of single use and hence find themselves forced to buy 'fresh' on a regular basis. For this is also the case with access to many leisure pursuits and services. Most tickets bought for entry to theatres, cinemas, exhibitions and concerts, as well as those for travel on buses, trains and planes, are 'single use' in so far as they are only valid for one occasion. Or, like most vouchers and gift certificates, have specified expiry dates, or alternatively, as with some travel cards and phone cards, require 'topping up' on a regular basis if they are to continue to be valid. Finally, there is a whole swathe of permits and licences (e.g. passports and driving and gun licences) that also require renewing if they too are to continue to allow the consumer access to the service provided. One might think that, when it comes to the question of the drain on the Earth's resources caused by the consumption of disposable products, train tickets, MoT (Ministry of Transport) certificates, passports and the like hardly count. Yet resources are used up here too and if single-use tickets were replaced by multiple-use ones, or licenses renewed less often, savings would be made.

The reference to expiry dates serves as a reminder that a crucial part of the ideology underpinning this form of replacement consumption is to be found in the twin concepts of 'the fresh' and 'the stale'.[5] Fresh products need to be purchased not simply because the previous purchases have been consumed, but because – although still in existence – their use-value has expired, or become 'timed out' as it were. This is the justification for the use-by dates, which can now be found attached to foodstuffs in most Western countries. In the United Kingdom, these were introduced in the 1990s and their use is monitored by the government's Food Standards Agency.[6] The rationale for the introduction of such dates is clear enough given that, in the case of some products, there can be a danger to health if these are kept for too long before being consumed.[7] However, it is a concept that can be applied to products other than foodstuffs and medicines, and hence employed to encourage consumers to purchase 'fresh' products even when the 'old' still contain utility.[8]

The crucial point here is that consumers do not necessarily engage in replacement purchasing of this kind because they have a preference for the new over the old but simply because once used for the purpose for which they were designed these products have no utility left and consequently have to be replaced. Technically this does not necessarily mean that all those products described as 'disposable' can only be used once (one may use a disposable ballpoint pen on more than one occasion before it runs out), but rather that the product is sold on the understanding that it has limited utility and hence a short life.[9] Consequently, consumers usually buy these products knowing full well that they are not just replacing an item, but rather that they are engaged in a continuous or serial replacement process.

Of course, it is possible that a preference for the new does influence purchasing behaviour in relation to replacement purchasing of this kind in those instances where a clear choice exists between a disposable and a non-disposable product, both of which serve to meet the same need. It is possible that some consumers might prefer to have a new but disposable product (for example, a camera or a razor), rather than a long-lasting and multiple-use one which, as a consequence, eventually begins to show the signs of wear. Another reason why consumers might prefer disposable products to their non-disposable alternatives is because these don't require the owner to spend either time or money in cleaning or maintaining them; nor, because of their relative lost cost, do they need to worry especially if the object concerned is lost or stolen.[10]

The worrying tendency here – from the point of view of curbing if not preventing over-consumption – is the marked increase in recent years in the number of products that fall into the category of disposable, or single-use, products. For although some goods of this kind have been around for a long time (paper plates date from at least the beginning of the twentieth century), there has been a significant growth in recent decades. Disposable razors and nappies, for example, date from the 1970s, while disposable cameras in their present form date from the 1980s, and one-day disposable contact lenses from the 1990s. Currently disposable video cameras and even cell-phones are becoming more common.[11] Why has there been this increase in disposable products? Price and convenience would appear to be the two major reasons why people buy disposable rather than their non-disposable equivalents, and it is very clear that the price of many of these products has dropped sharply in recent years.[12] For example, in the United States the average price of a cell-phone dropped from $600 in 1990 to $162 in 2001 (Lapoix, 2011). The same phone can be bought today for under $50. Not surprisingly, perhaps, some people now regard these as in effect 'disposable' products. Are there any signs that this trend for increasing numbers of disposable products to enter the market might be slowing down or restricted in any way? Well, there is of course the encouraging history of the single-use plastic shopping bag, and the fact that over the past decade there has been a worldwide movement toward either banning or restricting its use ('Plastic Shopping Bag', *Wikipedia*).

However, this would seem to be something of a special case, as there appears to be little prospect of effective legislation being passed in the immediate future in relation to the majority of disposable products. Hence, although environmentalists continue to encourage consumers to avoid purchasing disposable goods, while offering suggestions concerning alternatives (see *Ecosimply*, 2009; Lawrence, 2011; Smith, 2011), it seems probable that this trend will continue.

Out-of-use replacement consumption

When we move on to consider the second form of replacement consumption mentioned earlier, that which occurs when products break, or become damaged or faulty in some respects, and consequently have to be replaced, we encounter the phenomenon of planned obsolescence. This is because the rate at which replacement consumption of this kind occurs appears to be largely related to the degree to which products are made to survive many years of use, although this rate also depends on the cost of repair in relation to the price of replacement. Planned obsolescence is the deliberate designing of a product to have a limited useful life, the classic examples of which are the nylon stocking and the electric light bulb. The inevitable 'laddering' of stockings made consumers buy new ones, and for years discouraged manufacturers from looking for a fibre that did not ladder. Exactly the same was true of the light bulb, although in this case not only did long-lasting bulbs exist – although ignored by most manufacturers – but also bulbs were deliberately made fragile to shorten their useful life (Wong, 2012). In fact, planned obsolescence isn't just one thing, but rather 'a range of manufacturing and marketing techniques that all share a single aim: encouraging consumers to buy more and so keep factories busy and products flying off the shelves' (Lapoix, 2011; see also Howard, 2013).

Thus, while the most obvious technique employed by manufacturers to achieve this aim is to make products weaker, less durable or impossible to repair, they may also lobby for new legal requirements and standards that mean consumers have to buy a new product.[13] An example of this kind would be the UK government's MoT test for automobile safety, roadworthiness and exhaust emissions, introduced in 1960. Originally only cars that were ten years old or more had to be tested in this way, but this was changed in 1967 to cars that were three years old, while checks for emissions were introduced in 1994. At the same time, the number of items that have to be checked has steadily increased over the years, thus increasing the potential number of items that might need to be replaced.[14] Consumers may also be forced to replace a product not because it is broken or faulty but simply because a change in the operating or regulatory system has rendered it obsolete and consequently useless. An example of such a process would be the digital switchover, that is to say the process by which analogue television broadcasting is converted to and replaced by digital television. In examples of this kind, consumers are not simply being forced to buy a new product, they are being forced to buy a novel one. Despite these examples of regulatory prompted obsolescence, the primary form taken by planned obsolescence is still the manufacture, by producers, of products that are poorly made and likely to break down, and there are reasons for believing that this phenomenon has become more extensive in recent years.

Paradoxically, one reason why this might be the case is the invention of more durable and long-lasting materials. For although this has made it possible to extend the prospective life of many products it is not in the manufacturers' interests to do so, as this would impact negatively on the sales of their products. Consequently, it has been suggested that planned obsolescence is actually resorted to more today than would have been the case in the past as manufacturers seek to offset the potential loss of profits that would result from employing these materials to make products last longer. The second factor is the change in the relative balance of cost between repairing and replacing a product. While the costs involved in the manufacture of many goods have steadily fallen (mainly due to manufacture being relocated in areas of cheap labour), the costs associated with undertaking repairs have, by contrast, not fallen accordingly (mainly because repair tends to occur in areas of expensive labour). Consequently, given that manufacturers know from experience that consumers will calculate that it is cheaper to replace a broken product than repair it, there is little incentive for them to design an unbreakable product. Lastly, the shortening lifespan of products (see discussion below) means that there is simply little point in manufacturers designing them to last. More than this, it means that they have little incentive to spend time in product testing. Consequently, products are not so much designed to fail after a given period of use as simply not tested to assess their robustness or potential longevity in the first place. Indeed, as Prabha Gopinath, executive vice-president at TestQuest, a firm that creates testing software used by manufacturers such as Handspring, Palm, Motorola and Nokia, observes, 'No one that I know exhaustively tests anything that is built ... that goes for PDAs, cell-phones, any software that's out there' (quoted in Spencer, 2002). In view of this, it is hardly surprising that Darren Blum, a senior industrial engineer at Pentagram Design, a firm that builds portable devices and computers for companies like Hewlett-Packard, says 'We joke that we design landfills' (ibid.). This neglect of testing naturally follows from the rapid turnover of goods as technology companies focus their efforts on bringing out new products rather than helping customers retain their existing ones. It also follows that, given the short lifespan of the products they produce, they actually have little or no need to engage in the traditional form of planned obsolescence and may focus instead on taking steps to prevent customers from repairing their products (Kahney, 2011), or alternatively, on ensuring that there is no second-hand market for their goods.

The other crucial dimension relevant to the form of replacement consumption that follows from products being broken or faulty, relates to the question of guarantees and warranties. A warranty is a promise, by the manufacturer, that they will offer repairs or replacement of a faulty product, while a guarantee is more of a promise that the product or service in question will meet the customer's satisfaction. While a warranty will be valid only for a limited period, a guarantee could be open-ended although these too are, more usually, only good for a set period of time. The valid life of both of these has been getting shorter and shorter in recent years. In 2002, Dell Computers slashed their warranty period from three years to one, while Apple Computer's iPod digital-music player comes with only a 90-day warranty. More and more products, especially in the electrical and electronic sector, come with highly restricted warranties, for example, ones that only cover labour costs, or are only valid if purchasers pay extra for technical support (Spencer, 2002).

While planned obsolescence is generally regarded as a cynical ploy on the part of manufacturers to exploit the public, making them buy goods more often than they need to, this is not always a justifiable criticism. For the cost of manufacturing a product that would have an effective life of many decades might not only render it too expensive for most people to afford but would be a rather pointless exercise if customers are already in the habit of changing products every few years for fashion or style reasons. Having said this, there is evidence that some people are trying to fight back against manufacturers' efforts to force them to replace products more often than is necessary (Feldman, 2012). Most notable among them is Beneto Muros, the man who created an everlasting light bulb, and his Sin Obsolescencia Programada or SOP movement (Valero, 2013).

Before leaving this discussion of replacement consumption there are two important qualifications to make. First, consumers do not necessarily have to buy new products when engaged in purchasing fresh or replacement products. These can be obtained second-hand, even if consumers cannot by definition buy truly disposable products this way. However, as suggested, not only are manufacturers seeking ways to ensure that their products cannot be reused, but the trend towards shorter and shorter product life-cycles, more untested products coming on to the market and changes in legislation and regulation, are all working against the possibility of products being recycled in this way and therefore the possibility of replacements being obtained second-hand. Second, one can have replacement experiences that do not involve either any additional strain on the Earth's resources or the generation of waste. For example, one can dine at a new restaurant instead of one's usual eating establishment or, assuming that no additional travel is involved, holiday at a different destination. Having said all this, replacement consumption, whether of disposable products or those intended to last, may well in most cases be a form of consumption of the new that is forced on consumers rather than something that stems from a preference for the new. This is much more likely to be the case with the two other forms of consumption identified above, those that are innovation or fashion-driven.

Innovation-driven consumption of the new

In describing the second type of consumption of the new outlined above, that is to say consumption of what are seen as innovative goods, one is referring to the purchasing of products on the assumption that they offer a more efficient, convenient or satisfying manner of meeting the consumer's needs or wants. A large number of products of all kinds are frequently presented to the public in this fashion, that is as being either 'new' or 'improved', sometimes indeed as both 'new and improved', even though, as has been noted, strictly speaking this is a contradiction in terms (*Television Tropes and Idioms*, n.d.). In reality, it is relatively rare for genuinely new products to be launched on the market, new that is in the sense of constituting an invention that either finds a new means of satisfying an existing need or creates a means of satisfying a want that formerly did not exist. Examples might be the Sony Walkman portable audiocassette in the late 1970s or the GPS-based automotive satellite navigation system (Sat-Nav) in the 1990s. As both these examples suggest, genuinely new products usually stem from the development of new technologies, a process that would appear to have accelerated considerably in recent years. One, admittedly approximate, indication of this is the fact that increasing numbers of patents are filed worldwide every year; a patent being the registration of a machine or a process that, being non-obvious, qualifies as something that is either novel or an improvement on what went before (Gurry, 2008).

Yet it is not just the rate at which new technologies are developed that is critical in determining how quickly consumers take up genuinely new products; for this also depends on how quickly they become aware of their existence. But then inventions in communication technology have helped to speed up this process. The natural consequence is that the rate of adoption of new technologies is faster than it has ever been. For example, it took all of seventy years (from c1900 to c1975) before 90 per cent of people in the US had a telephone in their home. It has, however, taken only some twenty years (from c1985 to c2005) for 90 per cent of people in the US to possess a cell-phone (Mulbrandon, 2001).[15] This rapid take-up of new technologies necessarily means that the typical life-cycle of manufactured products has shortened dramatically in recent years. One measure of this is the fact that in the US at least half of annual company revenues across a whole range of industries now come from the sale of products that were launched within the past three years (Horn, 2013); while the average lifespan for white goods has fallen from 10 to 12 years before 2000 to just 6 to 8 or 9 years today (Lapoix, 2011). Essentially, as Jane Spencer of *The Wall Street Journal* puts it,

> The newer the product, the shorter the lifespan: A black and white TV sold in 1979 lasted for about 12 years; today a cutting-edge LCD-screen TV is replaced after five. Lap-top computers need to be fixed every 16 months on average, while hand-held organizers last an estimated two years.
>
> *(Spencer, 2002)*

As these last two example suggest, this rapid rate of turnover of products is especially marked in a field like electronics, where a company like Intel is working on the production of the next generation of PC chips before it has even begun to market the last one. But then this is also the case with pharmaceuticals, and in an area like nanotechnology, where it is estimated that, in the US, three to four new products come on to the market each week (*The Project on Emerging Nanotechnologies: News Archive*, 2008). Of course some of these may simply be replacements for old products, but then this still implies novelty even if the total number of products remains unchanged So, to summarise, more new products are coming on to the market, and at a faster and faster rate, and are being adopted by consumers quicker, than ever before.

The same thing is generally true of products that are best described as 'improved' rather than new. This is no better illustrated than in the example of the internal combustion engine, which has been consistently developed, over the 130 or so years since it was first invented, such that the latest models are all generally more powerful, quieter, more durable, less polluting, more efficient as well as safer, than their predecessors, a rate of improvement that also shows little sign of slowing down. On the other hand, it is worth noting that the vast majority of products that are designated as 'new' are not in any real sense new products; nor indeed do most of those described as 'improved' deserve this label. This is because in most countries government regulations merely require that there is a small functional change in a product or its packaging for it to qualify as 'new or improved'. As far as the majority of foodstuffs are concerned, including drinks, this criterion is easily met. All that is necessary is for the manufacturers to make a small alteration to the proportion of a given ingredient. A little more, or even less, salt or sugar, for example, would be sufficient to allow for such a label to be added to the product.[16] Much of this fake, or disingenuous improvement, appears to be driven by the manufacturer's fear that the consumer will get bored with 'the same old product', and may even defect to a rival producer if their own products are not regularly presented as 'new and improved'. Then, on top of this, there is also the perception that consumers have come to expect the products offered on sale in the marketplace to be continually 'improved', an expectation that suppliers feel obliged to try to meet. However, in cases such as these, one could claim that all the consumer is really consuming is the idea, or the impression, of the new or the improved, not the reality.

One crucial feature of this rapid rate of adoption of new technology, especially in a field like electronics, a feature that connects with the above discussion of planned obsolescence, is that existing products may quickly become unusable. Whereas in some fields products that embody old technologies can still be used alongside the new – for example, people who want to carry on using fountain pens rather than switch to biros can still do so – in other areas the new technology displaces the old in such a manner as to make the old effectively obsolete, often by driving the equipment necessary to use the old off the market; in the way that DVDs rendered videocassettes obsolete for example. This is particularly marked in the case of electronic goods where digital obsolescence, that is to say, 'a situation where a digital resource is no longer readable because the physical media, the reader required to read the media, the hardware, or the software that runs on it, is no longer available' ('Digital Obsolescence', *Wikipedia*) is widespread. This is becoming a more common occurrence and is even deliberately engineered by some producers. Thus, the computer software manufacturer, Microsoft, ensures that, having introduced a new file format, the older versions that consumers already possess are no longer supported. This practice is an example of what has been called systemic obsolescence, which is itself of course a sub-type of planned obsolescence.

Now while such techniques as these may well have the effect of forcing some consumers to buy products that embody 'new' or 'improved' technologies, it is also clear that there are many who need no such arm-twisting. These are the technophiles or gadget fans, people who are prepared to queue for up to fifty-three hours outside Apple stores in major cities round the world just to buy the latest iPad or iPhone the minute it goes on sale (*Mail Online*, 2011). For these people, owning the latest in the series of such devices gives them a 'new possession rush' (Bethel, 2013), and there is little doubt that they obtain real pleasure from being able to buy the latest high-tech devices. It should be noted, however, that the producers of these high-tech products do sometimes have a problem in matching the highly specialized but insistent demands of gadget fans with the needs of the less demanding but much larger audience for their products (Wilhelm, 2013).

Fashion-driven consumption of the new

The third type of consumption of the new outlined above referred to that form of consumption in which the driver is novelty rather than freshness or technical innovation, although of course the products concerned may also be both of these things. However, the dominant consideration here, as far as the consumer is concerned, is an aesthetic one, with taste the key criterion governing choice. However, just as has been noted in relation to the other forms, it is also the case that there has been a marked acceleration in this form of consumption in recent decades. This is most obvious in relation to clothing, although it is important to remember that fashion considerations apply to a very much larger range of retail products than apparel. When the Western fashion pattern first became established in the middle of the eighteenth century, the change in taste that marked the move from one fashion style to the next tended to occur, at the fastest, on an annual basis (Campbell, 2007a). However, after the Second World War fashions began to change twice a year as 'fashion weeks' were held in the principal Western capital cities of Paris, Milan, London and New York. These were occasions at which the major designers or couture houses could display their fashions for either the Autumn/Winter or Spring/Summer season. However, in recent decades increasing numbers of designers have been staging inter-seasonal collections between these two major fashion occasions, primarily to help shorten the customer's wait for a new season's clothes ('Fashion', *Wikipedia*). However, until recently the extent to which a fashion season could be shortened in this way was restricted by the basic fact that it took a minimum of four months to design, manufacture and then ship new apparel to the retail outlets where it was to be sold. Not any longer, however, for new technology, together with instant feedback from the stores concerning what is selling and what is not, as well as the increased involvement of customers themselves in this feedback process, has meant that it is now possible to bring this down to a matter of weeks. Thus, *Primark* boasts that, once a style has been identified, it can take as little as six weeks to reach the stores, while *Zara*, who have their own manufacturing and distribution division, can have catwalk-inspired items in stores within two weeks (*Ruck Retail Solutions*, 2012). For it is now the case, as Andrew Groves, course director for fashion at the University of Westminster, puts it, that 'We ... live in a fast-paced society. Pictures of what's on the catwalks of London Fashion Week today will be on the Internet today. Everything is absorbed quicker and we want it quicker. Looks hit the High Street much faster' (Winterman, 2009).

One natural consequence of this 'fast fashion' is that consumers buy more clothes than ever before, partly because clothing has become much cheaper over recent decades, and partly because what we could call the 'wardrobe life' of clothing has become so much shorter. An indication of how much cheaper clothing has become over the past forty to fifty years is that although the percentage of total expenditure that women in the UK spend on clothes has remained roughly the same (at around 5 to 6 per cent), the total amount of clothing they have bought has soared. This is because clothes have effectively become some 70 per cent cheaper over that period, mainly because of the massive expansion of production in the Far East (Cohen, 2010). In fact, much of this acceleration has occurred over the past two decades, with one 2011 study in the UK revealing that the average woman bought half her body weight (c 62lb (c 28kg)) in clothing in a single year, which was four times as much as she did twenty years earlier (Sims, 2014). Naturally, this increase in clothing bought also means an increase in clothing thrown away, as consumers discard their older purchases to make room for their latest acquisitions (Shields, 2008).

The other factor that is relevant here is that, every year, more and more people are caught up in the process of following fashion. This is perhaps most obvious with respect to developing societies, where increasing affluence is making this possible. For example, China's fashion market is growing at a phenomenal speed, and it is estimated that it will account for 30 per cent of the global fashion market's growth by 2016, becoming the world's second biggest fashion market by 2020. Indeed, it is already the case that some young consumers in China (those under 25) are spending 40 per cent or more of their disposable income on fashion (Zhang, 2011). But then it is also the case that more people are being drawn into following fashion in the developed societies of the West, largely through the recruitment of sections of the population previously considered immune to its appeal. For example, people at both ends of the age range have increasingly begun to take an interest in fashion. Thus, women in their sixties and seventies, who in previous generations would generally have settled into a fixed mode of dress by that age, are now keen to wear the latest fashions, including the latest accessories. In the UK, this phenomenon has been called the 'Twiggy Effect' after the 1960s model who, at the age of 63, is the advertising face of Marks and Spencer. As the appropriately named Professor Julia Twigg sums it up, 'Women over 75 are now shopping for clothes more frequently than they did when they were young in the 1960s' (Cohen, 2010). But the appeal of fashion has also been extended at the other end of the age range, with teenage girls and even pre-teen girls becoming increasingly fashion conscious, using cosmetics at an earlier age and being influenced in their dress by pop stars and celebrities (Kapoor, 2008; Pilcher, 2011). Finally, there is also evidence to suggest that men are now a far more important part of the fashion market than they were only a few decades ago, with demand for menswear rising at a faster rate than that for womenswear, especially in China and the Far East (Thakur, 2011).

Summary

There has been a marked increase, in recent decades, in the total number of products coming on to the market, and within this an increase in both disposable products and products that are intended to have a restricted life. There has also, at the same time, been a marked shortening of the life-cycle of those products not normally regarded as disposable and consequently a greater turnover of goods and increase in the generation of waste. This is true not only of those products that are primarily of aesthetic significance but also of those of a technological nature, as digital obsolescence meets up with fast fashion. In addition, each year more people around the world become affluent enough to adopt this modern style of consuming, while more sectors of the population are also drawn in as traditional age and gender barriers are broken down. Viewed in this way, it becomes possible to see the current phenomenon of hyper-consumption as largely stemming from a marked increase in the consumption of the new in all its forms; that is as an increase in the consumption of products that are seen as fresh, those that are viewed as technologically innovative and those that are seen as fashionable.

In fact, it would appear that these particular distinctions are increasingly breaking down, such that consumption of the new becomes a blend of all three, with technological innovation blending with the idea of replacement consumption and the following of fashion, something that has effectively become institutionalised through a process by which manufacturers sequentially label their products. Thus products are known as 'first generation', 'second generation', etc. or even actually sold with numbers attached. Thus, the iPad 5 replaces the iPad 4 and Grand Theft Auto 5 replaces Grand Theft Auto 4. At the same time, the rapidity of these model changes suggests that it might not be long before the life-cycle of some technological products will be as short as that for any given style in fast fashion.[17] All of which gives rise to two crucial questions: what is driving this process of ever-increasing consumption of the new and, in view of the threat that it clearly poses to a truly sustainable society, can it be stopped, or at least, slowed down?

The causes of hyper-consumption

The above discussion suggests that the two factors which combine to account for this phenomenon are technological advance and the progressive reduction in the price of products. Basically, while technological advance has made it possible for a continuing supply of new products to be brought to the market in an ever-shorter time, plummeting prices have made it possible for consumers to buy them in ever-larger numbers. Indeed the two factors work to reinforce each other, for without low prices there would be no mass market for new products and without a mass market there is little incentive for manufacturers to invest in the research and development that would lead to their creation. Unfortunately, this explanation does not immediately suggest an obvious solution to the problem. However, it does show that it is clearly a mistake to see the problem of over-consumption stemming from the accelerating consumption of the new as primarily a problem of individual behaviour. For although some environmentalist, and indeed some social scientists, appear to believe that this problem could be solved if only individuals would curb their tendency to buy new things, this view is clearly naïve. New technology does not come into being simply because there are technophiles desperate to get their hands on every new gadget that comes on to the market. Nor do new fashions arise simply because some consumers are dedicated followers of fashion; nor, least likely of all, do products appear on supermarket shelves marked 'new and improved' simply because shoppers have become bored with the 'old and unimproved' items they were accustomed to purchase. In fact the apparently ever-increasing tendency to consume the new is neither simply down to insatiable consumers who are unable to control their preferences or compulsions, nor is it simply the product of unscrupulous manufacturers and marketers who so engineer and present their products that people have little choice but to buy them. To accord blame in this way to just one side of the demand and supply equation would be mistaken. For clearly both have their part to play in this process, with consumers' determination to up-grade and dispose of electronic goods or to embrace the latest trend or fashion equally matched by the manufacturers' determination to produce goods of a kind that, with an ever-shorter life-cycle will, of necessity, require continual replacement. To that extent, one could say that both parties to this process are locked together in an ever tighter institutional embrace. For although modern consumers could be said to crave novelty, with their preference for the new and the novel central to their capacity to generate new wants (Campbell, 2005), manufacturers are equally forced to plan for the new or improved if they are to remain competitive.

But then governments must also accept their share of responsibility for the process of introducing the new to the market since they determine the regulatory framework that governs the sale of goods, as well as the fiscal and monetary framework that guides much industrial research and development. Indeed governments, through various agencies, directly fund a considerable amount of the research that results in new technologies being developed and hence new products created, a good example of this being the numerous spin-offs from the work done by NASA in the USA (Jones, 2008).

Can turning to embrace the old slow the advance of the new?

This last example points directly to one of the obvious difficulties in suggesting that the processes identified above might be stopped, or even slowed down. For the rapid take-up of new technologies is largely driven by the advance of science, and yet it is hard to see that there would be much appetite for curbing this latter trend, even if it was thought practical. For the onward march of science clearly brings benefits, for example in a sphere such a medicine. Here there is a real expectation on behalf of the population at large that breakthroughs in treatment or the development of new drugs will continue much as has been the case for the past century or more. In that respect, it is obvious that no one wants the advance of scientific understanding to come to a halt. But also it seems unlikely that people would want progress in the general understanding of the workings of nature to slow down or stop. Indeed, it seems unlikely that there would be any appetite for foregoing the advances in understanding that science represents simply to slow the flow of new products on to the market. But then of course included in this flow are those very inventions that boost our general capacity for sustainability, such as low-energy light bulbs and solar panels, suggesting that an overall luddite-style rejection of technological advance would not seem sensible. Indeed, it is worth noting that were it possible to deny consumers access to this continuing flow of new products this would almost certainly have some negative consequences of its own.

There are not many real-life examples of what happens when consumers are denied access to the continuing flow of innovatory products, so the experience of

Cuban consumers when the United States instituted its embargo on exports to the island in 1962 is of some interest. Since the majority of cars in Cuba at the time were of American manufacture their owners were denied access to new models and hence had little alternative but to try to keep the existing cars on the road as long as possible. It is these 'Yank Tanks' as they are called, or old 1950s model Chevrolets, Fords and Dodges, that one can still encounter in Cuba to this day, often in use as taxis. Now on the one hand this seems like an excellent example of a sustainable practice, that is to say squeezing the very last drop of life out of a product rather than exchanging it for a new model every few years. In that respect, keeping a car on the road for over fifty years is markedly at odds with the behaviour of contemporary Americans who, on average, tend to replace their car every three years. Yet there is a high price to pay for such apparent sustainability, which is that these old Cuban cars, when judged against their modern equivalents, are very fuel inefficient, highly polluting and comparatively unsafe.

There is a similar difficulty with the suggestion that the flow of new fashions or aesthetically influenced products on to the market could be slowed down or even halted. Admittedly, there are recent trends that do seem, at first sight, to point in this direction, notably the move to embrace retro or vintage fashion, a move that is often seen as part of a larger eco-fashion movement.[18] Yet there are problems and paradoxes here too. For one thing, cost is something of an issue. Although some second-hand clothing may well be cheap, for genuine vintage items to survive they will have to have been well-made in the first place, something that results in the addition of a price premium; while the necessarily restricted supply of such garments also tends to push up the price when compared with modern, cheaply made and mass-produced items. What is more, the restricted supply has led some high street retailers to introduce their own 'vintage reference' or 'vintage style' items, that is to say brand new items of clothing that have been modelled on an earlier style of dress (*Miss Selfridge* advertisement, n.d.). What this tells us is that the trend to favour vintage clothing is itself a fashion or fad, and may well die away just as quickly as it arose (Fortune, 2012). Also, it logically follows that if buying vintage fashion is to become a well-established long-term trend, rather than merely a passing fad, then people must carry on buying new products.

Indeed, what we learn from these examples is that valuing the old over the new is not in fact any kind of solution to the current accelerating demand for new goods. As far as genuinely old items are concerned, what we might want to call antiques, the immediate consequence is simply an increase in their price given the inevitable restriction on supply. Then, as a natural response to this shortage, imitation antiques or reproductions are created to help meet demand; thereby making a new business out of the re-creation of the old. In addition, either the old is recent enough to lead to a bout of nostalgia, something that in turn creates a new market for versions of the old, as in re-makes of classic films or new versions of classic songs, or the old is so distant in time as to be experienced as exotic, that is to say as something that is novel or different. Consequently valorising the past would appear to be no solution to the problem of the ever-increasing consumption of the new.

How valuing the new is central to Western civilisation

We have already noted that science places a value on the new and novel. Indeed science could not advance without a continuing diet of new findings, even if genuine discoveries are relatively rare. But then this is also true of academe, which also depends on an endless diet of new data, new insights and new understandings. Indeed, it is worth remembering that the definition of a PhD is a 'Substantial body of original work undertaken by the candidate' ('Doctor of Philosophy', *Wikipedia*); and that ever more of these degrees are granted by universities round the world every day (Maslen, 2013). But then this marked valorisation of the new is equally true of the arts. Thus, whether one focuses on architecture, painting, literature, music, theatre or film, one encounters the same fascination with 'the latest thing' or whatever is judged to be 'ground-breaking' or generally *avant-garde*. But then this is just as true of popular culture, whether the focus is on books, films, songs or computer games, while intriguingly the most popular form of reading matter in modern Western societies is a genre called 'the novel', of which an estimated 100,000 or so are published, in English alone, every year (Wilkins, 2009). As the name of this genre suggests, the content of these is assumed to be new, in the sense of fresh, even when it could be said to be formulaic. Not surprisingly perhaps, the number of new words entering the English language every year is also accelerating (Atkins, 2013). But then so too is the number of films made each year (Armstrong, 2010), as well as the number of songs written and performed, and subsequently downloaded from iTunes (Etherington, 2003). Among this avalanche of new cultural items, commentators and media gatekeepers, as well as publishers, publicists and, of course, retailers, are continually on the lookout for 'the next big thing', whether it is the must-see film, the must-read novel, the must-hear group or individual singer, or, in a fashion context, the next must-have item.

But then we can hardly ignore the fact that all forms of the media, whether radio, television, newspapers, or indeed the various social networks available on the internet, are dominated by something called 'the news', that is to say by 'newly received or noteworthy information, especially of recent or important events' ('News', *Wikipedia*). This is a product that is also a good example of the contemporary addiction to novelty, with increasingly numbers of people so desperate to acquire the very latest information available through this continuous stream of digital data that the computer or smartphone has been called the supplier of electronic cocaine (Thompson, 2013). But then one tends to forget that here too, recent decades have seen an accelerated rate of consumption, following the arrival of 24-hour news in the 1980s and consequently TV channels devoted exclusively to providing what CNN call 'a non-stop, real-time, live news stream' (Cushion and Lewis, 2010). Finally, perhaps it is not surprising that, in a culture so obsessed with the new, there should be regular crazes or fads, a craze being 'an enthusiasm for a particular activity or object that typically achieves widespread but short-lived popularity' ('Craze', *Wikipedia*). These can be found in all areas of life, whether the focus is on health (as with food or diet crazes), dance, games or even toys. Their rise and rapid spread is, however, greatly assisted by the new digital media, with the result that planking is rapidly replaced by owling, owling by Tebowing, and Tebowing by Lion-King-ing (*Mail Online*, 2012).

What can be done about this accelerated consumption of the new?

As suggested earlier, it seems improbable that consumers, when acting on their own, could possibly halt this apparently out-of-control consumption of the new. Although this doesn't mean that all consumer resistance is pointless for, in individual cases, revolts may force manufacturers to modify their behaviour (Naughton, 2011). It is simply that such actions are unlikely to impact on the system as a whole. The same is true for manufacturers, who may also help by making products that last (Cooper, 2010; Laskow, 2012), or alternatively introduce trade-in schemes that help to reduce waste (although many such schemes are merely marketing ploys), though again neither practice is likely to make that much difference to the overall level of consumption of the new. One innovation that would immediately reduce the current high level of consumption of new products would be to make their price reflect the true cost of their manufacture, the true cost that is to the biodiversity of life on planet Earth, not to mention the quality of life of future generations. There is already some widespread concern about the low level of wages and unacceptable working conditions of those employees – mainly in the Far East – who comprise the work force tasked with making these products. But if, in addition to action to improve their wages and conditions, action was also taken to incorporate the true costs to the environment and the health and welfare of people alive today and in generations to come, then most of the products discussed above would no longer be so cheap as to be treated as either fully disposable, or merely the latest in a large series of anticipated purchases.

However, such action could only be taken by that crucial third party to this consumption of the new, which is to say by government; or, in this case, by the governments of the leading industrial nations of the world acting in unison. But then governments do already control markets, through organisations such as trading standards authorities, which exist to enforce consumer legislation, while they also regulate trading, and control competition, as well as influencing the activity of markets through the taxing of goods and services. In this way, governments exercise considerable control of what is sold and under what conditions, and even – to a degree – at what price. Whether the political will exists to carry through the reforms necessary to restrict or reduce the threat to a sustainable future that is posed by the accelerating consumption of the new is another matter. For this extreme valuation of the new and the novel would seem to be in the very DNA of Western civilisation (Campbell, 2007b). It is, it would seem, a basic characteristic of Westerners to seek out the new, the cutting-edge, the modish, the ultra-modern or the state of the art; to discover and explore, to delve into the unknown; and indeed to literally go where no man has gone before. It is thus both the source of our pride and, because of our apparent insatiable appetite for it, a curse, one that threatens to render life on this planet unsustainable. We can only hope that the compulsion to seek out the new also leads to new ways of escaping from the curse it has brought upon us.

Notes

1 There are other ways of conceiving of 'new' products. New to the market is one not considered here, as is legally new; there can also be new to the firm manufacturing them, as well as new to a given segment of the population. See silver surfers driving a rise in broadband up-take as an example of new to a segment of the population (Warman, 2013).
2 One trend that might be relevant here, and that could possibly have led to acceleration in the rate of consumption of this kind, is the increased incidence of divorce in most Western societies. For this often leads to couples who formerly shared possessions, including significantly a common home, to replace these with both separate accommodation and two sets of furnishings and living equipment.
3 It follows that there is little possibility, in relation to these products, of there being any reuse or indeed primary or secondary forms of recycling, although in some instances tertiary recycling might be possible. However, some people do try to re-use disposable products (Michael, 2012).
4 These are all products that are described as 'disposable' or 'single use' by producers, advertisers or retailers. It does not necessarily follow that consumers regard them as disposable. Alternatively, there may well be other products that consumers do regard as disposable, or generally treat as disposable, even though they are not advertised as such.
5 For a fascinating discussion of the concept of 'the fresh' and how it has become a central feature of the modern food production and distribution system, see Freidberg, 2010.
6 Sell-by dates by contrast were introduced primarily for the benefit of the retailers, and were not intended to influence consumers. These are now in the process of being phased out in the UK.
7 A roughly similar rationale accounts for the presence of the use-by dates found on medicines.
8 See Cruel, 2014, for an example of how a use-by date is applied to encourage consumers to buy new cosmetics.
9 In fact some disposable products may have a relatively long shelf-life as it were. For example an air filter may last for several months before it needs replacing.
10 It is even possible that a disposable product is chosen in place of a non-disposable alternative for environmental reasons. For example in 2005 the British Government issued a report that suggested that traditional, washable nappies (diapers) were no more environmentally friendly than disposable ones (Aumonier and Collins, 2005).
11 The disposable cell phone is printed on recycled paper and allows about 90 minutes of call time, which is outgoing only (Bellis, 2013).
12 Disposables are usually cheap, and although in many cases it might well be cheaper in the long run to buy a product that is designed to last rather than a series of single-use disposable ones the initial outlay may well deter some buyers.
13 Lift manufacturers succeeded in doing this in France after fatal accidents in Amiens and Strasbourg. The French government changed the law so that a huge safety-critical upgrade of the country's lifts was undertaken, costing between four and eight billion euros. However, there was nothing wrong with the design of the lifts. The accidents were caused by poor maintenance (Lapoix, 2011).

14 Current suggestions to lengthen the period before new cars need to be tested from the current three to four years has brought intense lobbying from the Retail Motor Industry who claim it will lead to job losses (*Motor Trader Magazine*, 2011).
15 A more recent example would be the website 'Twitter'. It took only three years, from 2008 to 2011, for the percentage of the US population who were aware of something called 'Twitter' to go from 5 to 92 per cent (*on digital marketing*, 2012).
16 There is the salutary case of the sanitary product sold under the trade name Tampax. Here the manufacturer brought on to the market an apparently 'new and improved' product, but one that included fewer tampons in a standard box. The consequence was that sales fell. As a result of this disappointing news, a 'new and improved' product was introduced to replace it. This, however, consisted of a box that contained the original number of tampons (*TV Tropes and Idioms*, n.d.).
17 There have been seven generations of iPhone between 2007 and 2013. So effectively a new version of this product comes out every ten months.
18 Since UK Oxfam launched its online shop in 2007, after noticing a heavy increase in people typing 'vintage' into the site's search bar, its online sales have increased by 400 per cent (Stewart, 2012).

References

Armstrong, E. M. (2010), '50,000 movies are made every year – is that too many?', *The Moving Arts Film Journal*, 27 May. Available online at http://www.themovingarts.com/50000-movies-are-made-every-year-is-that-too-many/ (accessed 20 September 2013).

Atkins, A. (2013), 'Tag Archives: how many words enter the English language each year', *Atkins Bookshelf*, 16 July. Available online at http://atkinsbookshelf.wordpress.com/tag/how-many-words-enter-the-english-language-each-year/ (accessed 16 August 2013).

Aumonier, S. and Collins, M. (2005), *Life Cycle Assessment of Disposable and Reusable Nappies in the UK*, Environment Agency. Available online at http://www.ahpma.co.uk/docs/LCA.pdf (accessed 26 September 2013).

Bellis, M. (2013), 'Disposable cell phone-phone-card-phone: Inventor Radice-Lisa Altschul creates the world's first disposable cell phone', About.Com Inventors. Available online at http://inventors.about.com/library/weekly/aa022801a.htm (accessed 20 September 2013).

Berry, R. G. (ed.) (2006), *Environmental Stewardship: Critical Perspective, Past and Present*, Edinburgh: T & T. Clark.

Bethel, J. (2013), 'The urge – a desire for the newest technology', Blog Clove, 4 August. Available online at http://blog.clove.co.uk/2013/08/04/the-urge/ (accessed 12 August 2013).

Campbell, C. (1992), 'The desire for the new: its nature and social location as presented in theories of fashion and modern consumerism', in Roger Silverman and Eric Hirsch (eds), *Consuming Technologies: Media and Information in Domestic Spaces*, London: Routledge, 48–64.

Campbell, C. (2005), *The Romantic Ethic and the Spirit of Modern Consumerism*, York: Alcuin Academic.

Campbell, C. (2007a), 'The modern Western fashion pattern, its functions and relationship to identity', in Ana Marta Gonzalez and Laura Bovone (eds), *Fashion and Identity: A Multidisciplinary Approach*, London: Berg, 9–22.

Campbell, C. (2007b), *The Easternization of The West*, London: Paradigm Publishers.

Cohen, T. (2010), 'Growing old stylishly: Twiggy effect proves age is no barrier for fashion fans', *Mail Online*, 17 March. Available online at http://www.dailymail.co.uk/femail/article-1258469/Growing-old-stylishly-Twiggy-effect-proves-age-barrier-fashion-fans.html (accessed 14 August 2013).

Cooper, T. (2010), *Longer Lasting Products: Alternatives to the Throwaway Society*, Aldershot: Gower.

'Craze', *Wikipedia: The Free Encyclopedia*. Available online at http://en.wikipedia.org/w/index.php?title=Special:Cite&page=Craze&id=597261176 (accessed 23 August 2013).

Cruel, J. C. (2014), 'Get organized this week-end and clean out your old make-up', *Popsugar*, 18 January. Available online at http://www.bellasugar.com/When-Throw-Makeup-Away-Guidelines-Cosmetic-Life-Span-1124422 (accessed 26 February 2014).

Cushion, S. and Lewis, J. (2010), *The Rise of 24-Hour News Television: Global Perspectives*, New York: Peter Lang Publishing.

Devall, B. (2001), *Deep Ecology: Living as if Nature Mattered*. Layton, UT: Gibbs Smith.

Devall, B. and Sessions, G. (1985), *Deep Ecology: Living as if Nature Mattered*, Salt Lake City: Gibbs Smith.

'Digital obsolescence', *Wikipedia: The Free Encyclopedia*. Available online at http://en.wikipedia.org/wiki/Digital_obsolescence (accessed 26 February 2013).

'Doctor of Philosophy', *Wikipedia: The Free Encyclopedia*. Available online at http://en.wikipedia.org/wiki/Doctor_of_Philosophy (accessed 26 August 2013).

Ecosimply (2009), 'Avoiding disposable products – what are the alternatives?' 28 October. Available online at http://ecosimply.com/avoiding-disposable-products-what-are-the-alternatives-993.html (accessed 22 September 2013).

Etherington, D. (2003), 'Charting the iTunes Store's path to 25 billion songs sold, 40 billion apps downloaded and beyond', *TechCrunch*, 6 February. Available online at http://techcrunch.com/2013/02/06/charting-the-itunes-stores-path-to-25-billion-songs-sold-40-billion-apps-downloaded-and-beyond/ (accessed 26 August 2013).

'Fashion', *Wikipedia: The Free Encyclopedia*. Available online at http://en.wikipedia.org/wiki/Fashion (accessed 10 August 2013).

Feldman, J. (2012), 'Apple's planned obsolescence: Customer revolt brews', *Information Week*, 22 June. Available online at http://www.informationweek.com/global-cio/interviews/apples-planned-obsolescence-customer-rev/240002583 (accessed 20 February 2013).

Fortune, J. (2012), 'Is vintage clothing passé?' *The Guardian*, 27 January. Available online at http://www.theguardian.com/fashion/fashion-blog/2012/jan/27/vintage-clothing-passe (accessed 16 August 2013).

Freidberg, S. (2010), *Fresh: A Perishable History*, Cambridge, MA: Belknap Press.

Gurry, F. (2008), *World Patent Report: A Statistical Review – 2008 edition*, WIPO. Available online at http://www.wipo.int/ipstats/en/statistics/patents/wipo_pub_931.html (accessed 26 February 2014).

Hartmut R. (2013), *Social Acceleration: A New Theory of Modernity*, New York: Columbia University Press.

Horn, K. (2013)'The product life cycle is in decline' *Opinion – outsourcing – Sourcing Focus*. Available online at http://www.sourcingfocus.com/site/opinionscomments/6458/ (accessed 27 February 2013).

Howard, B. C. (2013), 'Planned obsolescence, 8 products designed to fail: Manufacturers' planned obsolescence costs consumers and the environment', *The Daily Green*. Available online at http://www.thedailygreen.com/environmental-news/latest/planned-obsolescence-460210#slide-1 (accessed 19 September 2013).

Jones J. (ed.) (2008), 'NASA technologies benefit our lives', *NASA Spinoff: Office of the Chief Technologist*. Available online at http://spinoff.nasa.gov/Spinoff2008/tech_benefits.html (accessed 10 August 2013).

Kahney, L. (2011), "Is Apple guilty of planner obsolescence?" *Cult of Mac*, 20 January. Available online at http://www.cultofmac.com/77814/is-apple-guilty-of-planned-obsolescence/ (accessed 19 September 2013).

Kapoor, D. (2008), 'Young divas make fashion child's play!' *Times of India*, 22 June.

Lapoix, S. (2011), 'Planned obsolescence: How companies encourage hyperconsumption', *OWNI.EU: News, Augmented, Technology, Politics, Culture*, 9 May. Available online at http://owni.eu/2011/05/09/planned-obsolescence-how-companies-encourage-hyperconsumption/ (accessed 19 September 2013).

Laskow, S. (2012), 'Four innovations that will make the future less wasteful'. Available online at http://grost.org/list/four-innovations-that-will-make-the-future-less-wasteful/ (accessed 24 September 2013).

Lawrence, R. G. (2011), 'How to avoid disposable utensils and paper products', *Mother Earth News: The Original Guide to Living Wisely*, 7 June. Available online at http://www.motherearthnews.com/diy/tuesday-video-how-to-avoid-disposable-utensils-paper-products.aspx#axzz2ePXavcVX (accessed 23 September 2013).

Mail Online (2011), 'Dedicated followers of gadgets: Lucky few get the first prized consoles as iPad 2 goes on sale in the UK', 28 March. Available online at http://www.dailymail.co.uk/sciencetech/article-1369835/Thousands-queue-London-world-iPad-2-goes-sale.html (accessed 4 March 2013).

Mail Online (2012), 'Forget planking, owling and t-bowing ... The latest internet craze is Lion-Kinging', 4 February. Available online at http://www.dailymail.co.uk/news/article-2096416/Forget-planking-owling-Tebowing-The-latest-internet-craze-Lion-King-ing.html (accessed 26 August 2013).

Maslen, G. (2013), 'The changing PhD – Turning out millions of doctorates', *University World News* Issue 266, 3 April. Available online at http://www.universityworldnews.com/article.php?story=20130403121244660 (accessed 26 August 2013).

Michael, P. (2012), '21 disposable products you can re-use', *Wisebread: Living large on a small budget*, 14 March. Available online at http://www.wisebread.com/21-disposable-products-you-can-reuse (accessed 23 September 2013).

Miss Selfridge (n.d.), advertisement for 'Vintage Style'. Available online at http://www.missselfridge.com/en/msuk/category/clothing-299047/vintage-style-299070?geoip=noredirect (accessed 14 August 2013).

Moncrief, L. W. (1970), 'The cultural basis of our environmental crisis', *Science*, 170(3957): 508–512.

Motor Trader Magazine: Latest News (2011), 'Trade bodies lobby on MoT changes', 19 April. Available online at http://www.motortrader.com/latest-news/trade-bodies-lobby-mot/ (accessed 23 February 2014).

Mulbrandon, C. (2001), 'Adoption of new technologies since 1900', *Visualizing Economics*, 18 February. Available online at http://visualizingeconomics.com/blog/2008/02/18/adoption-of-new-technology-since-1900 (accessed 26 February 2013).

Naughton, J. (2011), 'Has the revolt begun against Apple's iPad app fees?' *The Guardian* 4 September. Available online at http://www.theguardian.com/technology/2011/sep/04/apple-ipad-apps-subscriptions-revolt (accessed 8 August 2014).

'News', *Wikipedia: The Free Encyclopedia*. Available online at http://en.wikipedia.org/wiki/News (accessed 24 August 2013).

on digital marketing (2012), '5 Factors that influence adoption rates', 25 January. Available online at http://ondigitalmarketing.com/textbook/foundations/factors-that-influence-technology-adoption-rate/ (accessed 26 February 2013).

Pilcher, J. (2011), 'No logo? Children's consumption of fashion', *Childhood*, 18(1): 128–141. Available online at http://chd.sagepub.com/content/18/1/128.abstract (accessed 12 August 2013).

'Plastic shopping bag', *Wikipedia:The Free Encyclopedia*. Available online at http://en.wikipedia.org/wiki/Plastic_shopping_bag (accessed 19 September 2013).

Ruck Retail Solutions (2012), 'From catwalk to high street: the fashion cycle', 19 September. Available online at http://ruckretailservices.blogspot.co.uk/2012/09/with-london-fashion-week-having-ended.html (accessed 10 August 2013).

Shields, R. (2008), 'The last word in disposable fashion', *The Independent*, 28 December. Available online at http://www.independent.co.uk/life-style/fashion/news/the-last-word-in-disposable-fashion-1213847.html (accessed 12 August 2013).

Sims, P. (2014), 'Britain's bulging closet: growth of "fast fashion" means women are buying HALF body weight in clothes each year', *Daily Mail*, 21 January. Available online at http://www.dailymail.co.uk/femail/article-1389786/Britains-bulging-closets-Growth-fast-fashion-means-women-buying-HALF-body-weight-clothes-year.html (accessed 12 August 2013).

Smith, M. (Veshengro) (2011), 'Avoid disposable utensils and paper products', *Green (Living) Review*. Available (online) at http://greenreview.blogspot.co.uk/2011/07/avoid-disposable-utensils-and-paper.html (accessed 23 September 2013).

Spencer, J. (2002), 'Companies slash warranties, rendering gadgets disposable', *Stay Free Magazine*, 16 July. Available online at http://www.stayfreemagazine.org/public/wsj-planned-obsolescence.html (accessed 23 February 2013).

Stewart, E. (2012), 'The rise of vintage fashion', *Urban Times*, 19 June. Available online at http://urbantimes.co/2012/06/the-rise-of-vintage-fashion/ (accessed 8 August 2014).

Thakur, M. (2011), 'Luxury menswear growing twice as fast as womenswear', *International Business Times*, 14 December. Available online at http://www.ibtimes.com/luxury-menswear-growing-twice-fast-womenswear-382886 (accessed 14 August 2013).

The Project on Emerging Nanotechnologies: News Archive (2008), 'New nanotech products hitting the market at the rate of 3-4 per week', 24 April. Available online at http://www.nanotechproject.org/news/archive/6697/ (accessed 27 February 2013).

Thompson, I. (2013), 'Computers are "electronic cocaine" that make you MANIC', *The Register*, 14 January. Available online at http://www.theregister.co.uk/2013/01/14/computers_electronic_cocaine/ (accessed 26 August 2013).

TV Tropes and Idioms (n.d.), 'Main/new and improved', TV Tropes Foundation. Available online at http://tvtropes.org/pmwiki/pmwiki.php/Main/NewAndImproved (accessed 22 February 2013).

Valero, M. (2013), 'The fight against consumerism and planned obsolescence: The everlasting light bulb', *Global Research News*, 30 May. Available online at http://www.globalresearch.ca/the-fight-against-consumerism-and-planned-obsolescence-the-everlasting-light-bulb/5336950 (accessed 20 February 2013).

Warman, M. (2010), 'Silver surfers driving web growth', *Daily Telegraph*, 8 December. Available online at http://www.telegraph.co.uk/technology/news/8186716/Silver-surfers-driving-web-growth.html (accessed 20 September 2013).

White, L. (1967), 'The historical roots of our ecological crisis', *Science*, New Series, 155(3767) (10 March): 1203–7.

Wilhelm, A. (2013), 'Microsoft tells Windows Phone fans to chill in response to complaints, more updates coming this year', *Tech Crunch*, 23 July. Available online at http://techcrunch.com/2013/07/23/microsoft-tells-windows-phone-fans-to-chill-in-response-to-complaints-more-updates-coming-this-year/ (accessed 12 August 2013).

Wilkins, M. (2009), 'How many new novels are published each year?', *Work Product: Research Notes in Fiction, Theory and Digital Humanities*, 14 October. Available online at http://mattwilkens.com/2009/10/14/how-many-novels-are-published-each-year/ (accessed 14 August 2013).

Winterman, D. (2009), 'The life cycle of a fashion trend', *BBC News Magazine*, 22 September. Available online at http://news.bbc.co.uk/1/hi/8262788.stm (accessed 10 August 2013).

Wong, C. (2012), 'Planned obsolescence: the light bulb conspiracy', *ESSA*, 12 September. Available online at http://economicstudents.com/2012/09/planned-obsolescence-the-light-bulb-conspiracy/ (accessed 24 September 2013).

Zhang, E. (2011), 'Fashion-hungry Chinese market expected to triple by 2020', *CNBC, Consumer Nation*, 22 July, http://www.cnbc.com/id/43826546 (accessed 14 August 2013).

3
THINKING WASTE SOCIOLOGICALLY

Paul Hewer

Heavy weather

Troubled times: wars raging and warfare taking new urban forms; democratic inertia and the failure of the civil; religious and political antagonisms on the march and gaining traction; networked whistleblowers confronting informational dictatorships; lifestyle choices of protected affluence driving climate change; and spinning-out of such contexts of adversity, practices of market reshapers and related social theorizers busy at work building discursive empires of scholarly advantage. In this brief addition to the volume, a work dug out from an incessant urge to practise a form of rethinking, an attempt is made to chart emerging processes of recalibration at work within global mixes of the visible and valuable, wherein the knots of the ecological imaginary undo themselves and their dark discourses of exigency and imperative.

Social theorizing is currently febrile in its intensity and desire to re-attune itself to such hazardous times, fetishizing narrow-gauge 'culture' as an ever-ready panacea. Ulrich Beck's vision of a *World at Risk* (2009) perhaps best captures the current cultural moment and its structuring of feeling:

> World risk society forces us to recognize the plurality of the world which the national outlook could ignore. Global risks open up a moral and political space that can give rise to a civil culture of responsibility that transcends borders and conflicts. The traumatic experience that everyone is vulnerable and the resulting responsibility for others, also for the sake of one's own survival, are the two sides of the belief in world risk.
>
> *(Beck 2009, p.57)*

A view echoed by Zygmunt Bauman, for whom the writing is certainly on the wall:

> On a negatively globalized planet, all the most fundamental problems – the metaproblems conditioning the tackling of all other problems – are global, and being global they admit to no local solutions; there are not, and cannot be, local solutions to globally originated and globally invigorated problems.
>
> *(Bauman 2010, pp. 25–26)*

Beck hits the nail firmly on the head when he suggests that:

> The hardcore sociological question is: Where is the support for ecological changes supposed to come from, the support which in many cases would undermine their lifestyles, their consumption habits, their social status and life conditions in what are already very uncertain times? Or to put it in sociological terms: How can a kind of cosmopolitan solidarity across boundaries become real, a greening of societies, which is a prerequisite for the necessarily transnational politics of climate change?
>
> *(Beck 2010, p. 255)*

Central to such a greening will be a rethinking of consumption practices as embedded within and woven into the everyday and its assemblages of desire, routine and habit. Commenting on the context of climate change, Shove explores the burdens these assemblages place on social theory as the necessity to

> conceptualize and analyse processes of recruitment and defection: how do practices like those of showering on a daily basis capture us, their carriers? And how is it that people defect from others, like cycling or walking to work? These questions call for new ways of integrating micro, meso and macro levels of enquiry.
>
> *(Shove 2010, p. 283)*

It is therefore necessary to practise a form of analysis which does not simply overprioritize the micro or the macro but envisages how actions are enfolded and enframed within the orbit of larger forces to move beyond the construct of the consumer self as 'bounded, individualized, intentional, the locus of thought, action, and belief, the origin of its own actions, the beneficiary of a unique biography' (Rose 1998, p. 3). It is towards an understanding of how market-ing practices may be recalibrating the environmental imaginary that the chapter now turns.

Market reshaping and the spirit of sustainable invention

For Beck, management and marketing stand squarely in the dock; capitalism and its incessant urge for innovation and market opportunity are to blame for our current troubles:

> All past and present practical experiences in dealing with uncertainty can claim equal justification; however, for that very reason they do not offer a

solution to the resulting problems. More than that, key institutions of modernity, such as science, business and politics, which are supposed to guarantee rationality and security, are confronted with situations in which their apparatus no longer has any purchase and the basic principles of modernity no longer hold automatically. As a result, these institutions are being judged completely differently – no longer as trustees but as suspects. They are no longer seen as *managers* of risk, but also as *sources* of risk.

(Beck 2009, p. 54)

But then as solution providers and pedlars in myth, hope and transformation, marketers have not been shy in seizing the market opportunities which such a climate of uncertainty generates; offering up new philosophies of eco-consuming to curtail such doubts and offer a measure of hope. *Trendwatching.com*, a consultancy firm that tracks emergent cultural trends, are thus quick to spot the trend for 'Guilt-free consumption' and 'Eco-superior'. Products and branding are thus positioned to engage us in an intimate and persuasive conversation around ready-made solutions to such problems of uncertainty. Talk of sustainability has thus become the new language game, not the only game in town but an easy sell in hazardous times. From the hybrid to the DreamWorlds offered up in the advertising of 'new' cars which are advertised with their CO_2 and eco-credentials to the fore. Another example is how such concerns filter into games, a great way of reimagining the world as it is and imagining change, for example Top Trumps games like *eco-action* which aim to produce new forms of practical consciousness and awareness centred on an ethic of 'Play. Learn. Discover your ecoworld'.

Coupled with such attempts to recalibrate our understanding of consumption as other to its identity-status-driven form where the new is cherished for its power to inspire and innovate will be a heightened sensitivity to everyday routines. Such routines include showering (Shove 2010), staying warm (Jalas and Rinkenen 2013), sorting rubbish or domestic water consumption (Vannini and Taggart 2013). This trend is likely to witness increasing scrutiny of consumption practices and see them subjected to increasing metered surveillance and micro-management. The reinvention of frugality as an ethos of 'smart' consumption is slowly gaining traction in ways that 'downshifting' has been unable to because of its connotations of sacrifice and denial when set against the more immediately visible and available secular gratifications of easy (irresponsible) consumer lifestyling. Disengaged consumption, where chains of impact and import remain uncontested reveals myopia at the heart of everyday lifestyle practices of accountability. We argue that Miller's notion of shopping as 'making love' (1998) indicates the scale of disconnect operating at the level of the everyday domestic, where taken-for-granted often poses as essential. An imaginary that generates its own forward motion through essentialisms of making love manufactures its own justifications too. And while the warp and weft of everyday life draws upon assemblages of essentialisms and justifications, routines and habits, must-dos and lists of priority, it remains impervious to recalibration as long as the priority is on the individual or the sheltered therapeutic of a home over-protected from larger forces of change.

A less blasé attitude towards impact and import may prove increasingly pervasive, driven by the growing expression of climate cost through the media of economic imperative. Macro forces will engender increased concern around a rethinking of the desire to make easy the sometimes troublesome and troubling routine solutions to the trials of everyday. For example, when switching on a light or a heater will become a moment of reflexivity: a moment of heightened self-awareness, a moment of self-accounting, a moment of metered behaviour for the character of impact and import that such routine actions call into being. Possibly, the reflexive consumer lifestyle will problematize the autonomy and unspoken privilege of the individual actor as a buyer or even consumer, reframing the naïve autonomy that informs the idea of the self as a necessary site of buying privileges and market solutions. Such reflexive moments and their ontological uncertainties and self-accountings, suggest a technology of being that calls anew upon the politics of experience as ecological narcissisms: so much so that new passions may emerge centred on an eco-logic of cost-saving, energy reduction and the desire to desist from disengaged irresponsible consumption. *Treading lightly* will thus take on a new urgency and appeal, producing its own forms of standing out, its own local cultures of taste and conformity, with the 'home front' becoming the battleground for such a rethinking of how lifestyling may be better practised. In this climate, new forms of practical and eco-consciousness may surface, driven by a range of forces. One such force will be the shift towards new forms of metered existence. Here diagnostics will take centre-stage to assemble new platforms of performativity. Perhaps the mantra of the cover – reuse, recycle and refrain – appears more in tune and relevant in an era of global uncertainty and the responsibilities brought in the wake of such a realization. Such a vision of eco-Utopia will be better publicized, distributed and communicated through media platforms in search of an audience desiring to participate. Social media and TV programmes with their in-built reaching out and glamorizing potential will perhaps better underscore the necessity, or even the advantages to be claimed through recalibrating one's routines and habits. Through such cultural forms, change will be possible.

Closer to home we witness how the local re-sensitizes us to global problems. Take, for example, Greener Scotland, an organization set up with a transformative agenda of inspiring a population to turn to *living greener*. Here the call to arms is participatory, almost tribal, in its mantra of, 'Let's Go Greener Together':

> Greener living is a great way to build a cleaner and greener Scotland. What we've done so far has already had an effect, but we can do much more if we join together. We can have a huge impact on the wellbeing of our families, the comfort of our homes, the quality of the places we live, and the health of the natural environment around us. Greener living also helps us to play our part as one of the wealthiest countries in the world. We must all do what we can to reduce carbon emissions and slow down climate change. And if we look after the planet, we look after Scotland too. Cleaner air. Warmer

homes. Less noise. Less pollution. Better health and fitness. It all adds up to a better quality of life.

(http://www.greenerscotland.org/why-live-greener)

The website offers up a range of greener solutions around saving energy, greener travel, greener eating to translate the science of waste and reduction into a set of easy recipes for change. From advice on 'what can I recycle' to explaining the manifesto of reduce, reuse and recycle: 'By reducing waste, choosing goods that last and recycling those things we can't reuse or repair, we can all make a real difference to the environment we share' (www.greenerscotland.org.reduce-reuse-recycle). Here the discourse of *Greener Together* offers something akin to Connolly and Prothero's green consumption as life politics (as inspired by Giddens) where they suggest: 'The question of, "How shall I live?" has to be answered in day-to-day decisions about how to behave, what to wear and what to eat and numerous other things, as well as interpreted within the temporal unfolding of self-identity' (Connolly and Prothero 2008, p. 131). In this fashion, consumer culture appears to be reimagining itself with the everyday being rethought, old habits and routines disturbed and the reawakening of politics and ethics as central to contemporary consumption.

The tango of rationalization and romanticism

For Beck, Max Weber's *Protestant Ethic thesis* must be furthered, or as he suggests:

> According to Weber, the globalization of risk is not bound up with colonialism or imperialism and hence is not driven by fire and the sword. Rather it follows the path of the unforced force of the better argument. The triumphal procession of rationalization is based on the promised utility of risk and on the corresponding rational restriction of the side effects, uncertainties and dangers bound up with it ... However, the idea that precisely the unseen, unwanted, incalculable, unexpected, uncertain which is made permanent by risk, could become the source of unforeseen possibilities and threats that effectively place in question the idea of rational control – this is inconceivable on the Weberian model.
>
> *(Beck 2009, p. 17)*

The work of Colin Campbell (1987) comes to mind, with his incisive analysis of how ideals affect and are performed in practical conduct. Here the return to Campbell is necessary, for his thesis demonstrates that Weber's tale missed a vital ingredient, that the rendering of the iron cage of rationality and necessity masks a romantic spirit which dances to a different tune to that of strategy and calculation; a spirit best conjured in the desire to day-dream and fantasize. Is there a role for such self-illusory hedonism in a marketized culture of individual rage, narcissistic urge and immediate gratification? Whence comes anew the ecological imaginary of the frugal, of the treading lightly and related spaces of domestic practices of reflexivity and accounting: sidelined, erased, defeated by arguments of power and the status quo; or, do such forms of consumption simply go underground to resurface when most in demand. It is not difficult to see how musical forms will express such ideals. How the rhythm and moment of romanticism is best articulated through alternative forms of expression. The most obvious example would be music and its ability to conjure up a mood, a feeling, a context through sound. But advertising and branding, the twin tools of the marketing imaginary, sometimes provide alternative imagining solutions, spinning tales of hope and transformation that are hard to disavow. Salvation in an age of uncertainty and a world of ever-increasing risks appears to be centred on the shift to sustainable consumption. Such a shift, borne on the tides of many an advertising campaign for the latest eco-product, suggests that salvation may reside in the spirit of eco-logic; that it is through sustainable consumption and a rethinking of everyday routines that hope may lie. From Beck's (2009) questioning of rational thinking and the anticipation of risks as the force of current times to a recognition that while market practices may sometimes serve to heighten our sensitivity to such risks and problems, at other times such practices serve to quell and calm our sensitivity towards such risks offering a measure of amelioration to displace anxiety. The visual aesthetics of 'going green' express the iconography of sustainability and recycling as one of hope and transformation for commercial gain and personal adventure.

Waste and disposal

Waste announces itself through its facticity, through its sheer scale and the quantitativity of sheer numbers. In this game we are all entangled and complicit. Waste buttonholes us into political economy, the heavy-duty of social re-engineering and the mechanics of social reproduction; where culture squares up to the brute forces of economics, global politics and management.

Horizon 2020, the European Commission's new blueprint for climate action, captures and performs the emerging paradigm of discourses around 'waste' as a resource to recycle, reuse and recover raw materials:

> A smart economy minimises the production of waste and reuses waste as a resource. Resource constraints and environmental pressures will accelerate the transformation from a linear extraction-use-throw away model of production and consumption to a circular one. Moving towards a near-zero waste society not only has an environmental rationale, it increasingly becomes a factor of competitiveness. Europe has proven expertise in efficiently handling and treating waste and is at the forefront of innovation in this sector. Capitalising on these strengths, this call intends to further boost innovative, environmentally-friendly and cross-sectional waste prevention and management solutions.
>
> *(European Commission 2014)*

Here 'waste' becomes the great rescuer, the great hope, the means by which a troubled Europe will re-establish its position as 'world marker leader', for as the report opens: 'The global waste market, from collection to recycling, is estimated at EUR 400 billion per annum and holds significant potential for job creation' (ibid., p. 6). The old adage that there's *brass in muck* or even *every cloud has a silver lining* holds true. Or as the call to action (and grant salvation) continues: 'Industrial symbiosis, whereby different actors derive mutual benefit from sharing utilities and waste materials, requires large-scale systemic innovation with the aim of turning waste from one industry into useful feedstock for another one' (ibid., p. 8). The language of impact, innovation and action are to the fore with ambitious targets set to reduce food waste by 50 per cent by 2030, to reduce waste management costs and move towards 'sustainable food consumption patterns leading to healthier consumers and as a result reduced national health costs' (ibid., p. 11). A trip to the UK Government website for the Department of the Environment, Food and Rural Affairs reveals a similar tale of generating energy from waste with the management of waste as the new holy grail of governmentality: 'The UK is obliged under the revised EU waste Framework Directive to apply the waste hierarchy. This ranks waste management options in order of environmental preference and the first priority is waste reduction' (UK Government 2013).

What strikes the author is the absence of theorizing consumption from such policy-making debates of impact. Instead for the EU the talk is much more of the effort to model business and consumer behaviour through discussion of what are termed 'eco-innovative strategies', whereby proposals must highlight 'how urban patterns, drivers, consumer behaviour, lifestyles, culture, architecture and socio-economic issues can influence the metabolism of cities' (European Commission 2014, p. 16). What price hope and understanding in a world where modelling is to the fore; where quantitativity and the self-assurances it brings wins the day and better argument. Here as Bauman foretold 'Salvation is in numbers' (2004, p. 121). But what remains absent and erased is the recognition, which Miller (1995) in his discussion of *consumption as the vanguard of history*, so well foretold, that consumption remains first and foremost 'a social, cultural and moral project'. That is, an embedded assemblage of practices through which the pressures of politics and economics, the pressures of the everyday, of routines and habits are enfolded and speak us into being. Such a form of analysis brings with it an in-built sensitivity to life politics and to qualities over quantities. But more so, that it is only through understanding such shifting and fixed consumption practices that we glimpse how life unfolds in contested forms, how life flows in unintended ways, ever-responsive to shifting macro, micro and meso forces.

Such an acknowledgement that consumption, culture and lifestyle remain the crucial ingredients for a recipe for change and understanding is at least distinct from the talk of the *post-consumer* concept which one finds mobilized within the waste management literature. This is a significant field of academic inquiry which mainly hails from the heft and might of environmental sciences, engineering and manufacturing. With its own journals, favoured concepts and methodologies, the disconnect you witness here from consumer research startles, the disconnect from consumption as a social form is abrupt. One such concept deployed in this disciplinary field is that of the *post-consumer* and its attendant forms of *post-consumer waste*. Here *post-consumer* is defined in final, fixed and closed terms as simply end-of-life, no space here to acknowledge the social life of things (Appadurai, 2010). For example, Staikos and Rahimifard (2007) discuss trends in the global footwear industry which witnessed a 70 per cent increase in footwear production from 1990 to 2004, with worldwide footwear production and consumption doubling every 20 years from around 2.5 billion in the 1950s to 20 billion pairs in 2010. The challenge here is an environmental one; that is the waste generated in this post-consumer phase with most shoes ending up in landfills. But it is also a problem of design and product development; that is the different materials which are combined to produce a shoe fit for purpose and which subsequently become difficult to separate. Likewise Domina and Koch report on post-consumer textiles waste, 'garments or household articles made of textiles that the owner no longer needs and decides to discard' (Domina and Koch 1999, p. 347); revealing that most households recycle textiles with the most common form of disposition being to charity shops, family and friends or using as rags. While Ekström *et al.* (this volume) suggest that most clothes are not recycled or passed on in such ways but are destined for the dustbin.

Another example of such post-consumer modelling in action can be found in the work of Staikos and Rahimifard's waste management framework. This highlights a range of options for footwear from the proactivity of waste minimization achieved through design and material improvements, to reactive end-of-life management approaches with its focus on reuse, recycling, energy recovery and disposal. Here talk is of value recovery chains and the new market opportunities to develop through the market for recyclable materials (Staikos and Rahimifard 2007, p. 365). While for Huhtala, 'The percentage of waste recycled can be raised by increasing the participation rate of households in recycling programs and by increasing the number of waste items that can be reused, such as paper, aluminium, glass and plastic' (Huhtala 1997, p. 302).

The absence of theorizing consumption as practised here is troubling, not least from those academic diggers with a vested interest in such a topic, but concern also arises from the forms of knowledge produced around an understanding of waste when it is not adequately theorized through the lens of consumption. Here it is necessary to concur with Wilk (2002), for whom hope lies in identifying what he regards as the three major paradigms within consumption theory (individual choice, social theories of consumption and cultural theories of consumption) before outlining a Bourdieusian example based around theories of practice. Such an approach takes seriously notions of habitus, praxis and heterodoxy. Additionally, as part of this broadening of the conversation around consumption practices and their theorizing, it is useful to turn to work on disposal practices (Hetherington 2004; Munro 2013).

For example, Hetherington argues against the term waste for its negative implications of closure and lack of sensibility towards its 'dynamic and performative role within consumption' (Hetherington 2004, p. 159). Here the value of disposal reveals itself through rethinking the ways that people do 'membership and identity work' through not only practices of acquisition, but also 'how they dispose of consumed objects' (ibid., p. 167). Encouraging us then to think sociologically about disposal practices, Hetherington teases out the importance of 'how we manage absence – how we order it, place it, when we use it as a source of value' (ibid., p. 170). Here the focus is much more on the forms of organizing and order-building made possible through such value judgements. For Munro, much like Hetherington, 'Disposal is typically related in the literature – negatively – to waste . . . [but] the meanings of disposal in its everyday sense of material arrangement and placing are elided and, with this, much of the moral ordering of place and space is overlooked' (Munro 2013, pp. 214–215). Later in this insightful article, Munro likens his proposal to a rethinking of sociology:

> We need to do more than refine waste products as recyclable, as if we could re-instate the marginalized, and excluded back to their proper place in society. We need to find ways, as well, to challenge how it is that systems of production 'outcast' materials and morals as useless, outdated, or even immobile. It is not enough to try to 'balance out' the cast-offs of production and consumption perspectives by pointing to human rights, or by recourse to markets offering compensation.
>
> *(ibid., p. 225)*

Here disposal practices are re-theorized as the *placing and arranging of things* to better reveal their intimate link with the moral framing of our worlds. The morality of waste and recycling better attunes us to how such practices are part-and-parcel of the everyday and involve us in dilemmas over choice, value and oversight. Discourses of waste then unfold as ontological and social dilemmas to be acted upon or cast aside as insignificant or vital to the assembling of ourselves (see Rose 1998). So the opportunity and need for greater conversations between disciplines such as the social sciences and mechanical and chemical engineering certainly looks like an increasing necessity, given the shared concerns and contrasting modes and styles of theorizing, an opportunity which the discussions in Borås in 2013 valiantly kick-started.

Conclusion

And so it is here, that the digging of this chapter closes, not with statistics, or the brute forces of economic endeavouring but with leftovers and the flickering of hope. To practise that is a form of theorizing around consumption and its consequences as inspired by Miller (1998) and Bauman (2004). For as Bauman and May suggest: 'Sociological thinking, as an antifixating power, is therefore a power in its own right. It renders flexible what may have been the oppressive fixity of social relations and in so doing opens up a world of possibilities' (Bauman and

May 2001, p. 11). A paper dug out then from contexts of adversity, cultivated to practise a form of thinking and theorizing around notions of waste and disposal which better captures the spirit of reinvention and recycling at work. And so it is here, to poetics as a form of rethinking the unforeseen and the unintended that the paper closes.

By way of illustration, let us return to Bauman on *Wasted Lives: Modernity and Its Outcasts*. In this illuminating work around notions of waste and societing, he offers up two images of designs around waste, those of mining and farming (see Bauman 2004, pp. 20–22). Two designs and recipes for living offered up by the social theorist busy at work theorizing and labouring. For Bauman, farming speaks of continuity; whereas mining speaks of rupture and discontinuity. But perhaps we should add a third image to these compelling designs, modes of organization and forms of order-building, that of gardening. To garden speaks less of mining or farming, less of rupture but instead of labours, remembrance and futures possible, on the *here and now* and the *what could be*. To garden speaks of continuity, but it expresses alternative forms of production, less tied to notions of strategy, calculation and profit. To garden does not demand a balance sheet of equities and taxes. To garden is a moral practice with in-built social tendencies. To garden is to assemble, to place and to arrange things. To garden is to be in tune with the passing of time, to sensitize oneself to collective endeavour through acts of caring and doing. To garden speaks of practices of digging and composting, of value as expressed in forms of making-do and doing. In the garden, 'waste' is recycled and reused. Recycling expresses itself through routine and habit. In the garden, death, decomposition and tragedy are never far. In the garden, the unforeseen and the unintended happen through happenstance as much as design.

For a glimpse into such practices of garden making at work, I urge you to take a look at the gardens of the homeless as documented in Balmori and Morton's *Transitory Gardens, Uprooted Lives* where they suggest:

> In the reuse of nearly everything discarded, the sparing use of water, and the economical treatment of space, these gardens speak the language of our time. We are admonished to recycle, to conserve, to make maximum use of scarce of natural resources. Here all of these admonishments are heeded by necessity, under extreme conditions, and the result is the elevation of such things as water and living plants to precious, valued elements. Ironically, it is through this necessity that the suffering underlying it that a respect for natural resources has emerged in these gardens.
>
> *(Balmori and Morton 1993, p. 7)*

To garden is then a form of politics borne of the desire for worlding and reassembling oneself and others. To garden speaks of power and affluence, but equally of vulnerabilities and forms of exclusion and inclusion. To garden is then to open oneself up to a world of possibilities; a world of hope, colour, beauty and transformation. In the garden we dwell.

Acknowledgements

This chapter could not have been written without the guidance and inspiration of Professor Douglas Brownlie. The author would also like to thank Professor Karin Ekström for her generous invitation to join the conversation and the stellar participants at the University of Borås workshop (Sweden) for their spirit, energy and knowledge on this subject.

References

Appadurai, Arjun (2010, orig. 1986), 'Commodities and the politics of value', in Arjun Appadurai (ed.), *The Social Life of Things: Commodities in cultural perspective*, Cambridge: Cambridge University Press, 3–63.
Balmori, Diana and Morton, Margaret (1993), *Transitory Gardens, Uprooted Lives*, New Haven, CT: Yale University Press.
Bauman, Zygmunt (2010), *Liquid Times: Living in an age of uncertainty*, Cambridge: Polity.
Bauman, Zygmunt (2004), *Wasted Lives: Modernity and its outcasts*, Cambridge: Polity.
Bauman, Zygmunt and May, Tim (2001), *Thinking Sociologically*, Oxford: Blackwell.
Beck, Ulrich (2010), 'Climate for change, or how to create green modernity', *Theory, Culture and Society*, 27(2–3), 254–266.
Beck, Ulrich (2009), *World at Risk*, Cambridge: Polity.
Campbell, Colin (1987), *The Romantic Ethic and the Spirit of Modern Consumerism*, Oxford: Basil Blackwell.
Connolly, John and Prothero, Andrea (2008), 'Green consumption: life-politics, risk and contradictions', *Journal of Consumer Culture*, 8(1), 117–145.
Domina, Tanya and Koch, Kathryn (1999), 'Consumer reuse and recycling of post-consumer textile waste', *Journal of Fashion Marketing and Management*, 3(4), 346–359.
European Commission (2014), "Climate action, environment, resource efficiency and raw materials, Horizon 2020 Work Programme 2014–2015. Available online at http://ec.europa.eu/research/participants/data/ref/h2020/wp/2014_2015/main/h2020-wp1415-climate_en.pdf (accessed 21 August 2014).
Hetherington, Kevin (2004), 'Secondhandedness: consumption, disposal, and absent presence', *Environment and Planning D: Society and Space*, 22(1), 157–173.
Huhtala, Anni (1997), 'Post-consumer waste management model for determining optimal levels of recycling and landfilling', *Environmental and Resource Economics*, 10, 301–314.
Jalas, Mikko and Rinkinen, Jenny (2013), 'Stacking wood and staying warm: Time, temporality and housework around domestic heating systems', *Journal of Consumer Culture*, published online 11 November 2013. Available online at http://joc.sagepub.com/content/early/2013/11/11/1469540513509639.abstract (accessed 8 August 2014).
Miller, Daniel (1998), *A Theory of Shopping*, Cambridge: Polity Press.
Miller, Daniel (1995), 'Consumption as the vanguard of history: A polemic by way of an introduction', in Daniel Miller (ed.), *Acknowledging Consumption: A review of new studies*, Oxford: Routledge, 1–57.
Munro, Rolland (2013), 'The disposal of place: Facing modernity in the kitchen-diner', *The Sociological Review*, 60(2), 212–231.
Rose, Nikolas (1998), *Inventing Our Selves: Psychology, power and personhood*, Cambridge: Cambridge University Press.
Sahakian, Marlyne and Wilhite, Harold (2013), 'Making practice theory predictable: Towards more sustainable forms of consumption', *Journal of Consumer Culture*, 14, 25–44 published

online 9 October 2013. Available online at http://joc.sagepub.com/content/14/1/25.abstract (accessed 8 August 2014).

Shove, Elizabeth (2010), 'Social theory and climate change', *Theory, Culture and Society*, 27(2–3), 277–288.

Staikos, T. and Rahimifard, S. (2007), 'Post-consumer waste management issues in the footwear industry', *Proceedings of the Institution of Mechanical Engineers, Part B: Journal of Engineering Manufacture*, 221, 363–368.

UK Government (2013), 'Generating energy from waste, including anaerobic digestion'. Available online at www.gov.uk/generating-energy-from-waste-including-anaerobic-digestion (accessed 8 August 2014).

Vannini, Phillip and Taggart, Jonathan (2013), 'Onerous consumption: The alternative hedonism of off-grid domestic water use', *Journal of Consumer Culture*, published online 11 November 2013. Available online at http://joc.sagepub.com/content/early/2013/11/11/1469540513509642.abstract (accessed 8 August 2014).

Wilk, Richard (2002), 'Consumption, human needs, and global environmental change', *Global Environmental Change*, 12, 5–13.

PART II
Managing waste

4
FACTORS AFFECTING DEVELOPMENT OF WASTE MANAGEMENT

Experiences from different cultures

Mohammad J. Taherzadeh and Karthik Rajendran

Introduction

The increasing global population with more than seven billion[1] people and improving global welfare has resulted in an increased daily consumption of resources and raw materials (Troschinetz and Mihelcic 2009). More than 2.5 billion tonnes of municipal solid waste (MSW) is generated per year in the world. Figure 4.1 shows the MSW generation per capita per day in 63 different countries (Ngoc and Schnitzer 2009; Troschinetz and Mihelcic 2009). Most of the developed countries produce more wastes than developing countries, and this is connected to the socio-economic status of the countries. In the global context, not everyone is interested in knowing what happens to their waste after it is thrown away. Over time, consumer research has paid attention to purchases. There is now a need to recognise the consumption cycle and to also consider waste (De Coverly *et al.* 2008).

The accumulation of waste and the 'throw-away-philosophy' has resulted in several environmental problems, health issues and safety hazards and prevented sustainable development in terms of resource recovery and recycling of waste materials. Local governments usually have primary responsibility for solid waste materials, and carry out their activities based on the primary concern with public health issues, and not 'environmental health'. Here, environmental health refers to the sustainability of waste handling and processing. In this context, effective removal of waste from neighbourhood residential areas has been prioritised, resulting in it being disposed of in sites outside the city boundaries, often in landfills. The limits to this approach became increasingly clear in industrialised countries during the 1960s and 1970s, as consumption patterns began to lead to a sizeable increase in waste flows, where disposal went beyond the limits of social acceptability and the absorption capacity of local and global sinks.

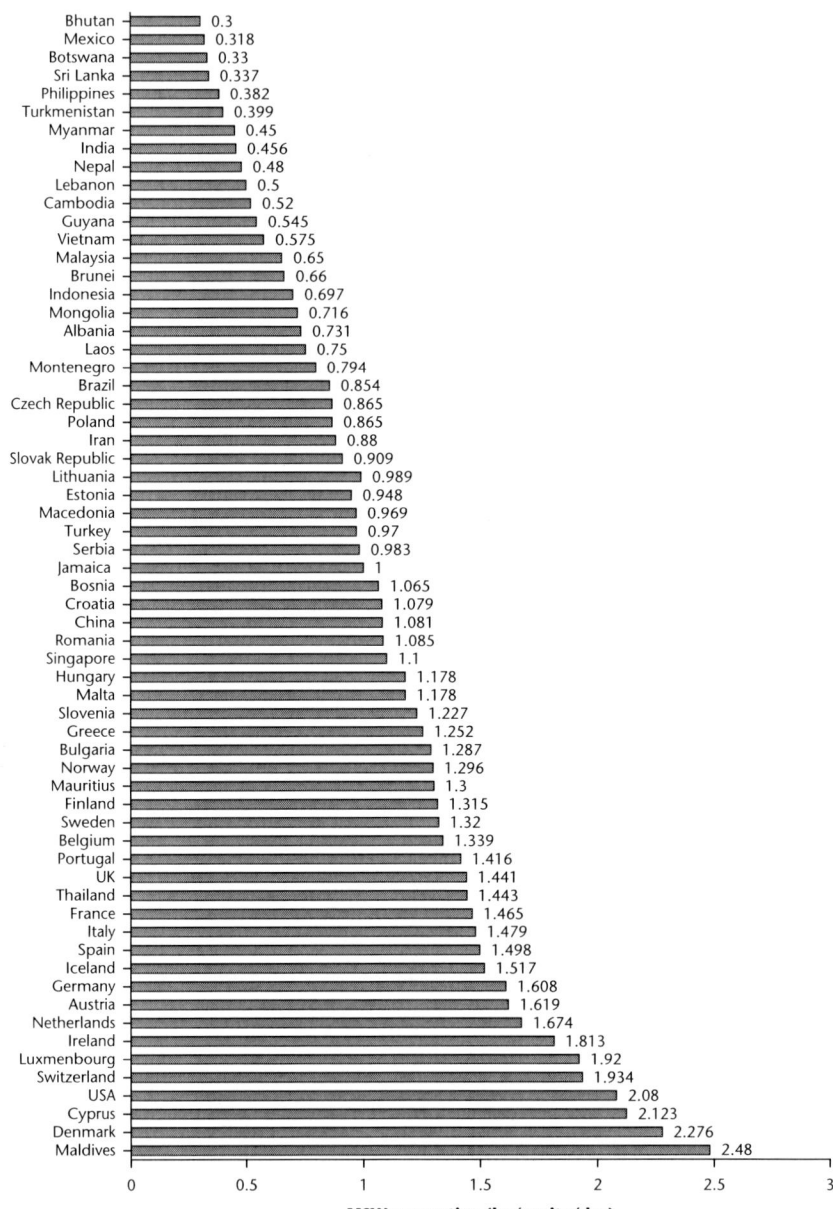

FIGURE 4.1 MSW generation rate in 63 different countries, in kg per capita per day

A perspective aimed at promoting greater sustainable development in the use of resources has influenced solid-waste management practices, and is gradually becoming implemented through policy guidelines at national levels in a number of industrialised countries. Guidelines and directives to reduce waste generation and promote waste recovery are established according to the 'waste management hierarchy', in which waste prevention, reuse, recycling and energy recovery are designed to minimise the amount of waste remaining for final, safe disposal (Rajendran *et al.* 2012). This discussion has been developed in some countries where old landfills are being restored and 'landfill mining' is considered a serious issue. Landfill mining is a process by which the old landfills are excavated and processed. However, from a global perspective, waste dumping or landfilling is still the major practice in several parts of the world. Other countries, such as Denmark and Germany, talk about zero waste, resource recovery and restoring the old landfills.

This chapter, which is mainly based on the main author's international experiences in Resource Recovery, aims to highlight different factors affecting waste management and resource recovery in a variety of countries. The ten countries studied have been selected based on many different factors, such as GDP, education, transparency index, social behaviour, policy planning, corruption, appropriate technology, waste collection, characterisation and separation techniques, religions, market for recycled material and people's awareness of sustainable cities. The countries are described briefly below before discussing different factors affecting waste management.

Waste management in different countries

The countries studied are categorised according to the extent of waste management practised. Group one consists of Sweden, Denmark and Japan, countries that have less than 3 per cent landfill. Group two consists of the United States, Indonesia, Qatar and India, countries that have between 69 and over 80 per cent landfill. Group 3 consists of Nigeria, Brazil and Romania, countries that have over 90 and up to 99 per cent landfill. Table 4.1 shows the waste management activities carried out by these different countries based on their waste generation.

Sweden

Basic facts

Sweden, with 9.5 million inhabitants, has seen great developments in waste management. In 1975, more than 1.6 million tonnes of MSW was landfilled in Sweden, while in 2012, this amount had reduced to less than 32,000 tonnes, which is a significant achievement (AvfallSverige 2011, 2012). In 2012, Sweden generated around 4.4 million tonnes of MSW or 460 kg/capita, of which 51 per cent of the waste was converted into heat and electricity through combustion. Sweden is located in the cold Nordic region, so heating is required for several months during the winter, while air conditioning is necessary for the shopping centres and hospitals during the summer. About 15 per cent of the MSW generated was sent to biological treatment plants in 2012 to produce mainly biogas, which was upgraded

TABLE 4.1 Different countries and their waste management activities

No	Countries	GDP (USD per capita/year)	Population (millions)	MSW generation (million tonnes/year)	Organics in MSW (%)	Recycling (%)	Incineration (%)	Compost/biological treatment (%)	Landfilling (%)
1	Sweden	43,180	9.6	4.4	35	34	50	15	<1
2	India	3,876	1,234	90	40–60			8	>80
3	Nigeria	2,661	173	25	50–60	8			>90
4	USA	49,965	316	389	25		7		69.7
5	Japan	35,178	127	51	26	18	74	5	<3
6	Indonesia	4,956	237	55	55	15		3	69
7	Denmark	42,086	5.6	9.5	40	24	53	18	<1
8	Brazil	11,909	201	95	52.5	1.2			96
9	Romania	16,518	20	21					>99
10	Qatar	98,329	1.9	2	57			15	80

to fuel for cars and buses (AvfallSverige 2012). Approximately, 32 per cent of the MSW was recycled, and 0.7 per cent went to landfills.

Waste management activities

As mentioned above, more than 99 per cent of the waste generated in the country is recovered or recycled, leaving no burden on the environment. This proper utilisation of the waste has been possible by changing the waste production patterns, implementing legislation since 1960s and designing less material for packaging (Hultman and Corvellec 2012). One of the main reasons behind the success of the Swedish waste management system is the waste separation at the source. This means that citizens separate the waste at their residence into about 30 parts according to their municipality model, which is an economic and effective way to manage waste. The more the waste is separated, the more value it has, as it can be easily recycled or converted into other materials or energy. Thus, the option of landfilling can be avoided.

Citizens are encouraged to separate the waste in two ways: one is by educating them about the waste and its importance and the other is providing economic incentives. For instance, a deposit system called PANT was created to increase recycling of PET (polyethylene terephthalate) and glass bottles and aluminium cans. The customers pay a deposit of 1–2 Swedish Krona (SEK) for each can or bottle, which is returned when they return the bottles or cans to any shop or supermarket in Sweden. As a result, more than 88 per cent of these materials are recycled in a relatively pure form (Rajendran *et al.* 2013). In Sweden, the governmental policies were formulated to enhance the extended producer responsibility (EPR) for packaging, waste paper, refrigerators, printers, etc. The EPR shifts the responsibility to the producer to collect it after its use and dispose of it properly (Pires *et al.* 2011).

Denmark

Basic facts

Denmark is also one of the Nordic European countries with many similarities to Sweden. The population of Denmark is about 5.5 million. However, Denmark, with 718 kg/capita/year in 2011, is at the top of the list in the EU regarding waste production. Nevertheless, the country has been able to manage it efficiently, with less than 3 per cent of its waste ending up in landfills. In 1993, about 9.5 million tonnes of waste was generated in Denmark of which 26 per cent ended up in landfills (Hjelmar 1996). The landfilling fee was constantly increased throughout the years from 160 Danish Krone (DKK)/tonne in 1993 to 355 DKK/tonne in 1999 (Arup Veltzé 1999; Hjelmar 1996). Since 1993, landfill has been reduced by 23 per cent landfill throughout the country (Larsen *et al.* 2010).

Waste management activities

Denmark is similar to Sweden in waste management activities, where the fee for waste collected is based on the polluter-pays principle. Additionally, there is also the PANT system and EPR for packaging, waste paper and cardboard (Pires *et al.* 2011). Denmark initiated an increase in its recycling activities in 1994–2005, which resulted in an increase in the recycling rate from 14 to 18 per cent. Out of the wastes generated, 53 per cent is incinerated, 24 per cent is recycled and 18 per cent is composted. (Gentil *et al.* 2009). In 2011, 5 per cent of total Danish electricity and 20 per cent of total district heat was produced through incineration (Habib *et al.* 2013). Combustion was the major waste management strategy to reduce the waste, which is why the country has more than 32 combustion plants, incinerating 54 per cent of the MSW.

Japan

Basic facts

Japan, an emerging economy in the Asia-Pacific region, with a population of 127 million; it produces about 51 million tonnes of MSW every year. Japan is one of the few countries in the world that has managed to control its waste production over a couple of decades (Shekdar 2009). This feat was achieved based on the numerous laws formulated and followed by the Government of Japan and its people. Japan began its waste management activities back in 1900 with the formulation of the Dirt Removal Law, followed by the Public Cleansing Law in 1954 and the Disposal and Public Cleansing Law in 1970 (Tanaka 1999).

Waste management activities

Due to its geographical and historical position, Japan has limited area for landfilling, which pushed the country to find alternative methods of waste disposal. The oil crisis of 1973 also emphasised the energy recovery from waste. This resulted in the country having more than 1,320 incineration facilities throughout the country (Horio *et al.* 2009). Incineration is the most common method of waste disposal in Japan with 74 per cent, followed by recycling with 18 per cent, and thereafter composting (5 per cent) and landfilling (<3 per cent) (Giusti 2009). However, Japanese regulations limit energy recovery from incinerators (Takaoka *et al.* 2011). In addition, Japan has designed one of the world's most advanced systems to sort waste at the source for its MSW (Geng *et al.* 2010).

USA

Basic facts

The USA, one of the most industrialised countries in the Western world with a population of 314 million; it generated 389 million tonnes of MSW in 2008

(Van Haaren *et al.* 2010). The USA started its waste management activities back in 1895 in New York City. It began with a unit operations approach, control on waste management, collection and transportation, processing, incineration and landfilling. (Kollikkathara *et al.* 2009). In 2006, per capita waste generation was around 2.1 kg/day, which had increased from 0.25 kg/capita/day in 1916. This is a 740 per cent increase over a period of 90 years (Kollikkathara *et al.* 2009). The majority of the wastes generated ended up in landfills; this is because there is more land available for landfill (Eighmy and Kosson 1996). However, it should be noticed that a strong private sector is mainly responsible for the waste management in this country and it is resistant to any change.

Waste management activities

The major parts of waste in the USA include paper products (34 per cent), food scraps and yard trimming (25 per cent), plastics (12 per cent) and metals (8 per cent) (Subramanian 2000). Mandatory recycling was incorporated in New Jersey in 1987, and by 1994 about 44.5 per cent of glass was recycled (Subramanian 2000). It was estimated that collection and transportation of waste costs more than the revenues generated from waste through recyclables and disposal costs (Kollikkathara *et al.* 2009). Today, about 115 waste-to-energy plants are in operation in the USA, and about 26 million tonnes of MSW are used in waste-to-energy projects. This has resulted in 2,700 MW of electricity production and 0.64 million tonnes of ferrous and non-ferrous metals recovery every year (Psomopoulos *et al.* 2009; Themelis and Kaufman 2004). In 2008, there were about 1,908 landfills throughout USA handling 269.7 million tonnes of MSW. However, the number of landfills is increasing (Van Haaren *et al.* 2010).

Indonesia

Basic facts

Indonesia is the fourth largest country by population, and it generates between 0.6 and 0.7 kg/capita/day of MSW (Meidiana and Gamse 2010). Since 1987, waste generation in this country has increased by 40 per cent (Maniatis *et al.* 1987). Waste is collected by three different methods: (a) door-to-door collection for each household, (b) jali-jali collection, through which a truck plays jali music so the people come out from their residences and deposit their wastes in the truck and (c) private collection systems. Approximately, 30 per cent of the waste is collected through private collection companies which collect the waste from residential and industrial areas (Pasang *et al.* 2007). About 80 per cent of the waste generated in Jakarta, the capital city, is collected and transported to a landfill, where 69 per cent of the waste is landfilled (Meidiana and Gamse 2010).

Waste management activities

Scavengers play a major role in the Indonesian waste management system. For instance, in the small town of Pontianak, with 550,000 inhabitants, there are 5,000 scavengers active in waste recycling. They take any valuables or recyclables from the wastes and sell it on the market. About 15 per cent of the waste generated is reduced through scavengers (Sicular 1992). Composting is also one of the upcoming methods, where more than 15 composting facilities, which can treat between 25 and 50 tonnes/day have been installed (Pasang et al. 2007; Zurbrügg et al. 2012). Moreover, 21 small incinerators are available with a processing capacity of 22 tonnes/day. The main barriers behind the waste management activities in Indonesia include policy barriers, technical barriers and lack of prospects of sustainability (Yao et al. 2011). In recent years, biogas production from wastes (e.g. the wastes from fruit and vegetable markets) to be used for electricity production is getting more attention.

Qatar

Basic facts

Qatar is a small country with an area of less than 11,000 km^2 and population of 1.9 million. It generates about 5,000 tonnes/day or 960 kg/year/capita of MSW. Qatar is one of the countries in the world with the highest GDP/capita. The country is small in area with a high density of population, so landfilling not an option. However, today more than 80 per cent of the waste generated ends up in landfills (Al-Maaded et al. 2012; Alhumoud 2005; Alhumoud et al. 2004).

Waste management activities

The wastes from households are packed in plastic bags and thrown in containers. From the containers, about 700 trucks collect the waste from 54,000 collection points and transfer it to the dumping station. The dumping station is located about 7 km outside the city. The waste is not separated until landfilled. However, some materials are recovered through scavengers in the landfills. Currently, there are no landfill regulations, which is why Qatar does not have any sanitary landfills. Due to this increase in waste generation, the country has started to concentrate on integrated waste management activities through private organisations because they can do it in a better way (Al-Maaded et al. 2012; Alhumoud 2005; Alhumoud et al. 2004).

India

Basic facts

India is one of the developing countries in the world with a population of more than 1.2 billion (Shekdar 2009). This country produces about 90 million tonnes of

MSW per year, which is expected to increase at a rate of 1.0–1.33 per cent annually per capita (Unnikrishnan and Singh 2010). Per capita waste generation varies between 0.1 and 0.5 kg/day, based on the urban and rural areas (Unnikrishnan and Singh 2010). On average, about 60–70 per cent of MSW generated is collected, while the rest is uncertain or there is a lack of information. MSW is usually collected from the households by waste sweepers and transferred to dust bins. From there, wastes are transported by trucks or manually to dumping sites. Previously, municipalities of the cities were solely responsible for the collection and transportation of wastes (Gupta et al. 1998). However, in recent years *door-to-door* collection of rubbish has been initiated and is handled by private companies.

Waste management activities

Separation of wastes is uncommon in India (Narayana 2009; Unnikrishnan and Singh 2010), while the major proportion of wastes is compostable materials (40–60 per cent) with a calorific value between 800 and 1,000 kcal/kg. This number can be compared with the calorific value of, for example, petrol that is about 11,000 kcal/kg. Calorific value is the amount of heat released by a substance, when it is combusted. Mumbai, the biggest city in India generates more than 6,000 tonnes of waste every day (Sharholy et al. 2008). The common methods of waste disposal or treatment include dumping (unsanitary landfills), and aerobic composting or vermicomposting (composting through worms). Although millions of small-scale bio-digesters exist in India, industrial plants to treat household wastes are not popular, due to the mixing of the wastes (Rajendran et al. 2012). However, some incinerators and bio-methanation projects have recently been gaining popularity (Unnikrishnan and Singh 2010).

Nigeria
Basic facts

Nigeria, in West Africa, is the seventh largest country by population in the world with over 170 million inhabitants. The population growth is increasing at a rate of 2.38 per cent every year, leading to an increase in the utilisation of resources, leaving a lot of rubbish. (Ogwueleka 2009). About 25 million tonnes of MSW is generated every year in this country, which gives a density between 280 and 370 kg/m^3. Lagos, the largest city in Nigeria, generates about 4 million tonnes MSW every year (Ishola et al. 2013; Kofoworola 2007; Solomon 2009). The waste comprises 50–60 per cent organics, and about 8 per cent of the waste is reused through scavengers after it is deposited in dumping sites or landfills.

Waste management activities

Only 50–70 per cent of the waste generated is collected, and about 77–95 per cent of the revenue is allocated to collection of waste and the rest is used for disposal

(Yusuf et al. 2008). In rural areas, waste is not collected properly; instead, people tend to throw it into open areas or burn it in their backyards, leading to air pollution. The wastes are usually collected in compactor trucks, side loaders, tippers and flatbed trucks, one to three times a week (Agunwamba et al. 1998; Imam et al. 2008; Solomon 2009). Currently, the government of Nigeria is looking at privatisation of the waste sector for collection, transportation and processing of wastes, as the municipalities have not been efficient enough (Uwadiegwu and Chukwu 2013).

Brazil

Basic facts

Brazil, with more than 200 million inhabitants, is the most populated and largest country in Latin America; 47 per cent of the population of Latin America lives in Brazil. Although this country has an environmental profile and generates the majority of its electricity (81.7 per cent) from hydropower stations, there is negligible waste management throughout the country (Loureiro et al. 2013). Brazil generates 260,000 tonnes of solid waste every day, so the per capita waste generation exceeds 0.74 kg/day (Lino and Ismail 2012).

Waste management activities

Waste management in Brazil is closely related to the economic and industrial development in the country (Münnich et al. 2006). About 95 per cent of the generated wastes are collected (Lino and Ismail 2011), while more than 88 per cent of the wastes end up in open dumps, followed by 9 per cent in controlled landfills and 3 per cent dealt with following proper disposal methods (Bezama et al. 2007; Lino and Ismail 2012; Münnich et al. 2006).

Romania

Basic facts

Romania is one of the eastern European countries. It generates about 36,720 tonnes of animal and vegetable wastes every year (Lorenz et al. 2013). With a population of about 21 million, the per capita waste generation in the country is 1.08 kg/day (Lorenz et al. 2013). This country landfills 99 per cent of the wastes generated, which is among the top countries in the EU engaging in landfilling (Lorenz et al. 2013).

Waste management activities

In 1995, about 6.9 million tonnes of MSW were generated, which were deposited in more than 2,000 landfill sites (Apostol and Mihai 2012). Out of these 2,000 sites, only 450 had a proper license (Thomas 1999). Recently, NGOs and state-owned environmental companies together formed a group to create awareness among children and adults to dispose of wastes properly and manage wastes efficiently (Iojă et al. 2012).

Factors influencing waste management

The presentation of the ten different countries shows a wide variety of waste management in the world. Some countries such as Sweden, Denmark and Germany have greatly developed their waste management, reducing landfills and even restoring landfills, while other countries such as the USA, India, Brazil and Qatar still use landfilling as the major practice in their waste management. There are several factors affecting this, such as legislation, environmental, social, technological, health, market and economic aspects. All these aspects need to be analysed and considered to find a model for waste management that works for a country or city. Before major investments are made, it is also important to examine the options on a small scale, for example a part of the city, to see if the model fits in an economic and sustainable way. It needs to be recognised that if a waste management system works well in a specific country or city, it cannot automatically be copied to another country or city. The complexity of the different factors involved needs to be recognised. Figure 4.2 shows the roadmap to achieving zero waste. In many countries, one or several of these factors have led to a non-effective waste management system.

Government policy and finance

The policy of the government and fund allocations for the waste management sector play a vital role in creating zero-waste cities throughout the world. For instance, Sweden, Denmark and Japan have reached a position where they have negligible landfill mainly because of governmental policy and funds allocated for this sector. Various cities and provinces in India have, on the other hand, presented many plans for waste management, but have had difficulties executing the plans due to the implementation of the policy or laws, corruption and social factors.

Financial issues and tariffs for waste collections are other factors affecting waste management. In Sweden, for example, mixed waste, that is unsorted waste, has the highest collection tariff, while sorted waste such as papers, plastics, metals and batteries are free of charge to dispose of at household recycling centres.

Political issues and long-term planning

In most of the countries, local politicians acting in municipalities and city councils decide on the planning and implementations of waste management. Since proper waste management is a complex question, long-term planning is necessary. In some countries, for example Sweden, such decisions are made across party borders, and different political parties agree on a plan. However, in many countries, particularly developing countries, this type of agreement does not exist, resulting in changes in the

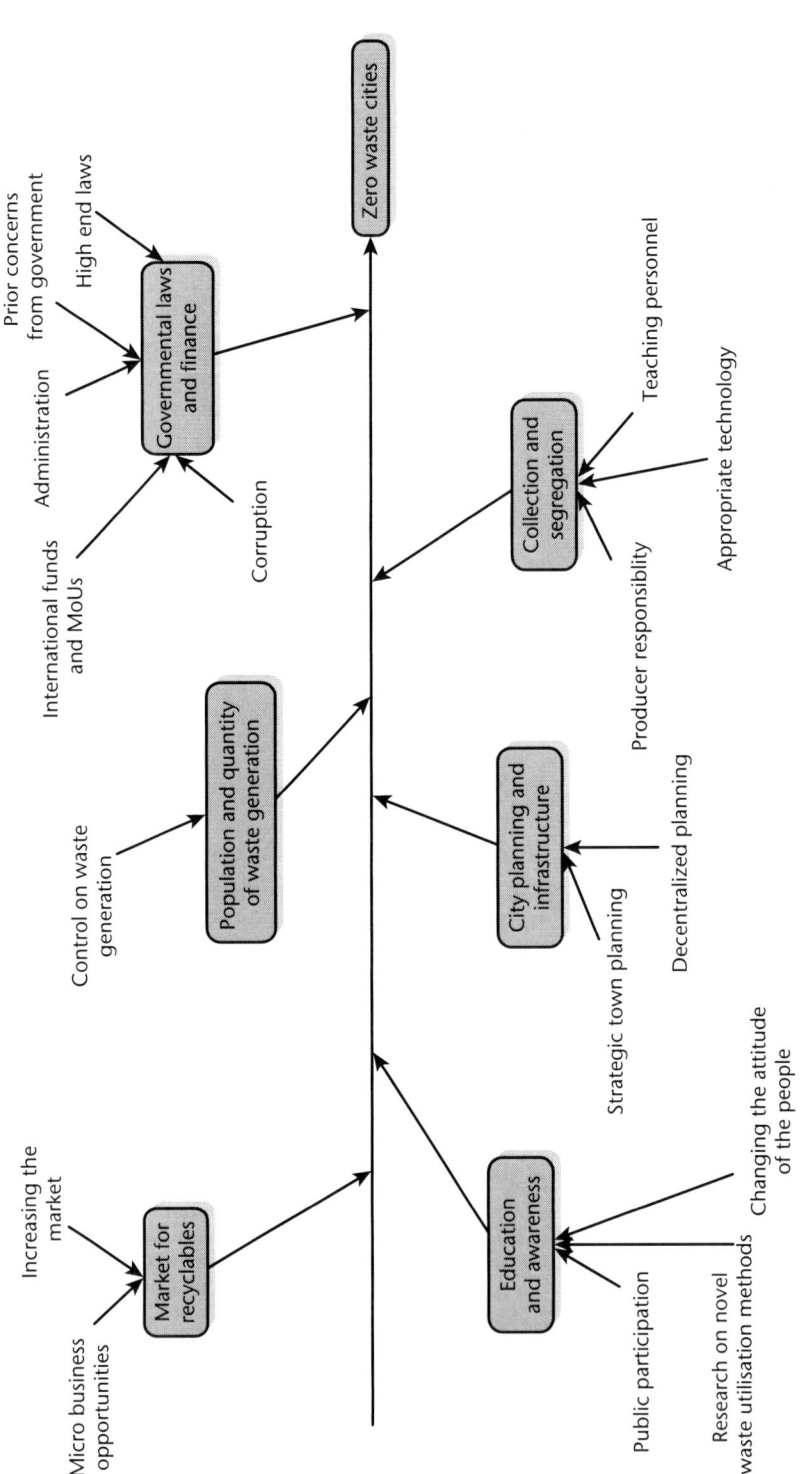

FIGURE 4.2 Roadmap to zero waste: factors, possibilities and challenges

plan following each election when a new party takes over. This is also valid for national plans; thus, to have more long-term planning, politicians should not change the waste plan each time a government changes, which is common in, for example, India.

Social behaviour in following the laws

When a new law and plan is enacted to alter waste management, the level of acceptance among citizens to adapt themselves to the new regulation is an important factor for the success of the plan. However, this acceptance is quite different in different countries because of different socio-cultural factors. For example, if a government enacts a law requiring batteries to be separated from the waste, some people may become suspicious of this new regulation and form conspiracy theories about it, while other people may voluntarily follow the new law.

Scavengers

Scavengers or non-organised private persons and companies are active in recycling valuable products such as metals, plastics or papers found in waste and, although illegal, they have built their own infrastructure in parallel to the legal waste management systems. They usually work under difficult conditions at dumping sites or stations and try to find and collect these materials and then sell it to relevant companies. One factor in changing the waste management to a modern system is to exclude scavengers. However, they are likely to be upset and afraid of losing their jobs and may try to resist such changes by sabotaging the plans. On the other hand, organising the scavengers is a difficult task because of the socio-cultural factors involved. One example of a solution, according to Professor Maria Cecilia Loschiavo at the University of São Paulo in Brazil, who conducted research in this field, is to integrate scavengers in waste recycling and taking care of collecting valuable wastes such as metals and plastics directly and legally from households.

Waste separation: machine vs human

Waste can be considered as a mixture of valuable materials. However, there is no technology to treat this mixture with an economically feasible process. Therefore, it is necessary to separate the materials in MSW before it is processed. There are currently two ways to deal with this separation: using machines or humans. Engineers have tried to develop a variety of waste-sorting machines and have installed them in different cities such as Bangkok, prior to, for example, composting. However, these machines are costly and none of them usually works properly. In low wage countries such as Thailand, the companies and municipalities try to complete this separation of mixed wastes by employing people to do it. The third way to deal with separation of waste is built on source separation by citizens. It means that citizens should not mix wastes, but should leave it at recycling stations (or trucks) already separated. If this works, there is no need to separate it afterwards. This method is much more effective, but relies on the collaboration of the inhabitants. Sweden has used this latter method, but decision-makers in many countries, such as India and some states in the USA, do not invest in the possibility of educating and changing the behaviour of the citizens.

Resistance to change

Any new method of waste management requires change and, therefore, resistance should be expected. It includes changing people's behaviour, new routines for companies and municipalities, and new investments. All these factors could inhibit changes to the system. Consumers often need environmental or economic motivations (e.g. Ölander and Thøgersen 1995), while companies sometimes, but not always, need to be forced by legislation. In addition, it should be clarified whether public or private investments will be used to finance the activity. Public investments usually need political motivations, while private investments demand profit. Furthermore, since the contract with companies in this branch is usually long, for example 20–30 years, this could be a hindrance to improving the system.

Waste compositions and the method of use

The methods for a proper waste treatment and resource recovery depend on the composition of the materials. For example, sorted wastes in Sweden contain about 3 MWh/tonne of energy and the country extracts about 2.5 MWh/tonne of energy in the form of heat and electricity via combustion and even extracts energy from burning hazardous wastes. The incinerators in Bangkok need to burn 100 kg LPG (liquefied petroleum gas) per tonne of sanitary waste. As the cost of LPG becomes too much, they cannot afford to burn all the household wastes. This means that Swedish companies extract energy and income from waste burning, while Thai companies have to expend energy and money for the burning. The fact here is that burning is only suitable for dry wastes and not wet mixed wastes. On the other hand, if the waste contains too much organic matter and water, then biological processes such as composting or biogas production could be more suitable. One such example is fruit wastes that are usually landfilled. In a joint effort, the University of Borås in Sweden and Gadjah Mada University in Yogyakarta (Indonesia) together with their municipality and companies built a biogas plant at the central fruit market in Yogyakarta. In this plant, the fruit wastes are digested to biogas and then converted to electricity to be used by the shops in the market.

Infrastructure and the need for products

There are different needs for products and different infrastructures in various cities, which are important factors in selecting the proper method for waste treatment. For example, Indonesia and Brazil have warm climates compared to countries such as Sweden or Germany. If the dry waste is combusted in these countries, Brazil and

Indonesia can probably benefit just from electricity production, as they principally do not use district heat. In this case, the investment might not be economically feasible. On the other hand, in colder climates such as Sweden and Germany, the combustion plants provide both heat and electricity to households and companies nearby. In this case, if pipelines to transfer the district heat were available (as they are in Sweden), a profitable investment could be obtained. However, if the investment for this pipeline is economically considered as part of the waste combustion plant, the economic feasibility of the plant might be difficult and the investment might not be attractive to investors.

Waste collection system

In all the industrial countries and even some of the developing countries, trucks carry waste collected from households to the treatment site or landfill. However, it is not the case in all countries and the transportation costs could be a bottleneck for selecting a waste treatment method. Also, building and running a large biogas, composting or combustion plant would be more economically attractive than small ones. However, the larger plants require waste collection from further away that is not possible everywhere. Consequently, the economy of large and small plants should be compared, together with the collection and transportation costs. In addition, in some cities, there are not enough trucks to transport the waste, and people transport it manually using small carts. In such cases, a plant might be able to cover a radius of 500 m and not more, which means that a large plant is out of the question.

Market for recycled material

The market for second-hand products is quite different in different countries and for different materials. There are markets for second-hand products such as clothes or home appliances, although many people do not buy from such markets. Again, there are socio-cultural differences. In Sweden, charity organisations collect 20 per cent of old clothes (Jeihanipour 2011); however, only a small percentage of these clothes are of good enough quality to be sold in second-hand shops. On the other hand, it is possible to recycle the materials. For example, a textile made of cotton and polyester can be pretreated to separate pure fibres of the cotton (or cellulose) and polyesters for recycling (Jeihanipour 2011). Paper fibres that are also cellulose can be recycled up to seven times for use in different product forms until the fibres become too short. Other examples are glass and metals that could be recycled to make new products. Electronic waste is also an example that is recycled to extract mainly copper and gold.

Land availability

Land availability is a factor affecting waste management in the world. The countries or areas with a low population (per area) usually start with the idea that they have no shortage of land, and therefore there should be no problem with using landfill for waste management. On the other hand, in some regions with high population, there is no land available for the landfill, and neighbours may protest when attempts are made to create a new landfill.

Household education and economics

Household economics is an important factor for waste management. Poor families usually have a lower amount of waste and try to reuse and recycle as much as possible. For example, a Russian study (Ekström *et al.* 2001) showed that families altered clothes so that they could be kept and used. Waste generation is generally correlated with the GDP of the country. Studies have also shown that educated people pay more attention to the environment (Bartelings and Sterner 1999).

Religion

Religion can help or hinder the improvement of waste management. In principle, most religions encourage cleanliness, as can be observed in relatively clean churches, mosques and temples all over the world. If the religion has a strong power in a society, religious leaders can motivate people to change their behaviour to improve waste management. For example, a German NGO built small biogas plants in villages to collect toilet wastewater and produce biogas, which in turn went to kitchens in the villages through pipes as fuel for cooking food. However, as the inhabitants in the villages were Muslims, they said the gas was Haram (forbidden to use). This religious obstacle was probably impossible for the NGO in Germany to understand. Finally, the NGO explained the problem to an Indonesian imam who ate the food cooked using the biogas and convinced the people that it was acceptable. Thus, the imam served as a gatekeeper for introducing waste management.

Corruption

Since waste management generally demands public–private partnerships in different aspects, it is also sensitive to the level of corruption in the country (Ahmed and Ali 2004). If this partnership is poorly designed, efficiency is reduced and corruption can increase. Corruption increases the investment costs and reduces the rate of return for the investment, which is already low in this area. An example in this regard is Nigeria (Agunwamba 1998), where investments and operational costs of waste management are relatively high because of corruption. Table 4.2 shows the comparison of various social factors in different countries discussed above.

Level of consumption

Finally, the level of consumption and population growth also affect waste management. If consumption continues to increase as it has done in the Western world

TABLE 4.2 Rating of social factors affecting waste management in different countries

Factors\Countries	Sweden	India	Nigeria	USA	Japan	Indonesia	Denmark	Brazil	Romania	Qatar
Economics	4	2	1	4	4	2	4	3	3	5
Transparency index	5	2	1	4	4	1	5	3	3	4
Population	5	1	3	2	3	2	5	2	4	5
Technology	5	3	1	5	5	2	5	3	3	5
Laws and policies	5	2	1	3	5	2	5	3	1	2
People participation	5	2	1	3	5	3	5	2	1	1
Education and research	5	2	1	5	5	2	5	3	2	3
Educating children	5	2	1	3	5	2	5	2	1	3
Market for recyclables	5	3	2	3	5	3	5	2	1	1

★ Scales: 1 – very poor, 2 – poor, 3 – good, 4 – very good and 5 – excellent

during the last few decades, we can expect that demand for waste management will continue to increase. For example, the consumption of clothes and shoes increased by 53 per cent in Sweden during 1999–2009 (Roos 2010). A national survey from 2011 (Gustafsson and Ekström 2012) shows that 62 per cent of Swedes throw away usable clothes and that 46 per cent of Swedes sometimes do this. Another example of this point is China, where the waste generation per capita increased from 0.5 kg/day in 1980 to 0.98 kg/day in 2006 (Zhang *et al.* 2010). It can be compared with the GDP per capita in China, which increased from 309 to 2,070 USD in the same period. The population growth demands continuous investment in infrastructure in waste management, even if the MSW production per capita remains constant.

Conclusions

The chapter has described different legal, social, economic, technical, educational and religious factors affecting waste management and resource recovery in a variety of countries. The chapter illustrates the need for knowledge in a variety of areas to develop a waste management system that works. Interdisciplinary research is warranted where researchers engaged in the fields of technology, production, materials and consumption cross borders and share and develop knowledge in this field. The complexity involved in waste management needs to be recognised. Waste management is no longer a national problem, but a global problem that needs international cooperation to find better solutions.

Note

1 Throughout this chapter, billion means 10^9 or a thousand million.

References

Agunwamba, J. C. (1998), 'Solid waste management in Nigeria: Problems and issues', *Environmental Management*, 22(6), 849–56.

Agunwamba, J. C., O. K. Ukpai and I. C. Onyebuenyi (1998), 'Solid waste management in Onitsha, Nigeria', *Waste Management & Research*, 16(1), 23–31.

Ahmed, Shafiul Azam and Mansoor Ali (2004), 'Partnerships for solid waste management in developing countries: Linking theories to realities', *Habitat International*, 28(3), 467–79.

Al-Maaded, M., N. K. Madi, Ramazan Kahraman, A. Hodzic and N. G. Ozerkan (2012), 'An overview of solid waste management and plastic recycling in Qatar', *Journal of Polymers and the Environment*, 20(1), 186–94.

Alhumoud, Jasem M. (2005), 'Municipal solid waste recycling in the Gulf Co-Operation Council States', *Resources, Conservation and Recycling*, 45(2), 142–58.

Alhumoud, Jasem M., Ibrahim Al-Ghusain and Hamad Al-Hasawi (2004), 'Management of recycling in the Gulf Co-Operation Council States', *Waste Management*, 24(6), 551–62.

Apostol, Liviu and Florin-Constantin Mihai (2012), 'Rural waste management: Challenges and issues in Romania', *Present Environment and Sustainable Development*, 6, 105–14.

Arup Veltzé, Suzanne (1999), 'Waste management in Denmark', *Waste Management and Research*, 17(2), 78–9.

AvfallSverige (2011), 'Swedish Waste Managment', Stockholm, Sweden.
—— (2012), 'Hushållsavfall I Siffror-Kommun Och Länsstatistik 2011'.
Bartelings, Heleen and Thomas Sterner (1999), 'Household waste management in a Swedish municipality: Determinants of waste disposal, recycling and composting', *Environmental and Resource Economics*, 13(4), 473–91.
Bezama, Alberto, Pablo Aguayo, Odorico Konrad, Rodrigo Navia and Karl E. Lorber (2007), 'Investigations on mechanical biological treatment of waste in South America: Towards more sustainable MSW management strategies', *Waste Management*, 27(2), 228–37.
De Coverly, Edd, Pierre McDonagh, Lisa O'Malley and Maurice Patterson (2008), 'Hidden mountain the social avoidance of waste', *Journal of Macromarketing*, 28(3), 289–303.
Eighmy, T. Taylor and David S. Kosson (1996), 'USA national overview on waste management', *Waste Management*, 16(5), 361–6.
Ekström, Karin M., Marianne P. Ekström and Helena Shanahan (2001), 'Families in the transforming Russian society: Observations from visits to families in Novgorod the Great', in A Gröppel-Klein and F.-R. Esch (eds), *European Advances in Consumer Research*, Vol. 5, Provo, Utah: Association for Consumer Research, 145–54.
Geng, Yong, Fujita Tsuyoshi and Xudong Chen (2010), 'Evaluation of innovative municipal solid waste management through urban symbiosis: A case study of Kawasaki', *Journal of Cleaner Production*, 18(10), 993–1000.
Gentil, Emmanuel, Julie Clavreul and Thomas H. Christensen (2009), 'Global warming factor of municipal solid waste management in Europe', *Waste Management & Research*, 27(9), 850–60.
Giusti, Lorenzo (2009), 'A review of waste management practices and their impact on human health', *Waste Management*, 29(8), 2227–39.
Gupta, Shuchi, Krishna Mohan, Rajkumar Prasad, Sujata Gupta and Arun Kansal (1998), 'Solid waste management in India: Options and opportunities', *Resources, Conservation and Recycling*, 24(2), 137–54.
Gustafsson, E and K.M. Ekström (2012), 'Ett Växande Klädberg', I L. Weibull, H. Oscarsson and A. Bergström (eds.), Framtidens Skugga, Som-Undersökningen 2011, Göteborgs Universitet, Som-Institutet: Som-Rapport No. 56.
Habib, Komal, Jannick H. Schmidt and Per Christensen (2013), 'A historical perspective of global warming potential from municipal solid waste management', *Waste Management*, 33(9), 1926–33.
Hjelmar, Ole (1996), 'Waste management in Denmark', *Waste Management*, 16(5), 389–94.
Horio, M., S. Shigeto and M. Shiga (2009), 'Evaluation of energy recovery and CO_2 reduction potential in Japan through integrated waste and utility management', *Waste Management*, 29(7), 2195–202.
Hultman, Johan and Hervé Corvellec (2012), 'The European waste hierarchy: From the sociomateriality of waste to a politics of consumption', *Environment and Planning-Part A*, 44(10), 2413.
Imam, A., B. Mohammed, D. C. Wilson and C. R. Cheeseman (2008), 'Solid waste management in Abuja, Nigeria', *Waste Management*, 28(2), 468–72.
Iojă, Cristian Ioan, Diana Andreea Onose, Simona Raluca Grădinaru and Cătălina Şerban (2012), 'Waste management in public educational institutions of Bucharest City, Romania', *Procedia Environmental Sciences*, 14, 71–8.
Ishola, Mofoluwake M., Tomas Brandberg, Sikiru A. Sanni and Mohammad J. Taherzadeh (2013), 'Biofuels in Nigeria: A critical and strategic evaluation', *Renewable Energy*, 55, 554–60.
Jeihanipour, Azam (2011), *Waste Textiles Bioprocessing to Ethanol and Biogas*, Gothenburg: Chalmers University of Technology.

Kofoworola, O. F. (2007), 'Recovery and recycling practices in municipal solid waste management in Lagos, Nigeria', *Waste Management*, 27(9), 1139–43.

Kollikkathara, Naushad, Huan Feng and Eric Stern (2009), 'A purview of waste management evolution: Special emphasis on USA', *Waste Management*, 29(2), 974–85.

Larsen, Anna Warberg, Hanna Merrild, Jacob Møller and Thomas Højlund Christensen (2010), 'Waste collection systems for recyclables: An environmental and economic assessment for the municipality of Aarhus (Denmark)', *Waste Management*, 30(5), 744–54.

Lino, F. A. M. and K. A. R. Ismail (2011), 'Energy and environmental potential of solid waste in Brazil', *Energy Policy*, 39(6), 3496–502.

— (2012), 'Analysis of the potential of municipal solid waste in Brazil', *Environmental Development*, 4(12), 105–13.

Lorenz, Helge, Peter Fischer, Britt Schumacher and Philipp Adler (2013), 'Current Eu-27 technical potential of organic waste streams for biogas and energy production', *Waste Management*, 33(11), 2434–48.

Loureiro, S. M., E. L. L. Rovere and C. F. Mahler (2013), 'Analysis of potential for reducing emissions of greenhouse gases in municipal solid waste in Brazil, in the State and City of Rio De Janeiro', *Waste Management*, 33(5), 1302–12.

Maniatis, K., S. Vanhille, A. Martawijaya, A. Buekens and W. Verstraete (1987), 'Solid waste management in Indonesia: Status and potential', *Resources and Conservation*, 15(4), 277–90.

Meidiana, Christia and Thomas Gamse (2010), 'Development of waste management practices in Indonesia', *European Journal of Scientific Research*, 40(2), 199–210.

Münnich, K., C. F. Mahler and K. Fricke (2006), 'Pilot project of mechanical-biological treatment of waste in Brazil', *Waste Management*, 26(2), 150–7.

Narayana, Tapan (2009), 'Municipal solid waste management in India: From waste disposal to recovery of resources?', *Waste Management*, 29(3), 1163–6.

Ngoc, Uyen Nguyen and Hans Schnitzer (2009), 'Sustainable solutions for solid waste management in Southeast Asian countries', *Waste Management*, 29(6), 1982–95.

Ogwueleka, T. (2009), 'Municipal solid waste characteristics and management in Nigeria', *Iranian Journal of Environmental Health Science & Engineering*, 6(3), 173–80.

Ölander, Folke and John Thøgersen (1995), 'Understanding of consumer behaviour as a prerequisite for environmental protection', *Journal of Consumer Policy*, 18(4), 345–85.

Pasang, Haskarlianus, Graham A. Moore and Guntur Sitorus (2007), 'Neighbourhood-based waste management: A solution for solid waste problems in Jakarta, Indonesia', *Waste Management*, 27(12), 1924–38.

Pires, Ana, Graça Martinho and Ni-Bin Chang (2011), 'Solid waste management in European countries: A review of systems analysis techniques', *Journal of Environmental Management*, 92(4), 1033–50.

Psomopoulos, C. S., A. Bourka and Nickolas J. Themelis (2009), 'Waste-to-energy: A review of the status and benefits in USA', *Waste Management*, 29(5), 1718–24.

Rajendran, Karthik, Solmaz Aslanzadeh and Mohammad J. Taherzadeh (2012), 'Household biogas digesters – a review', *Energies*, 5(8), 2911–42.

Rajendran, Karthik, Hans Björk and Mohammad J. Taherzadeh (2013), 'Borås, a zero waste city in Sweden', *Journal of Development Management*, 1(1), 3–8.

Roos, John Magnus (2010), 'Konsumtionsrapporten 2010', Center for Consumer Research, School of Business, Economics and Law, University of Gothenburg.

Sharholy, Mufeed, Kafeel Ahmad, Gauhar Mahmood and R. C. Trivedi (2008), 'Municipal solid waste management in Indian cities – a review', *Waste Management*, 28(2), 459–67.

Shekdar, Ashok V. (2009), 'Sustainable solid waste management: An integrated approach for Asian countries', *Waste Management*, 29(4), 1438–48.

Sicular, Daniel T. (1992), *Scavengers, Recyclers, and Solutions for Solid Waste Management in Indonesia*, Center for Southeast Asia Studies, University of California at Berkeley.

Solomon, Ugwuh Uchechukwu (2009), 'The state of solid waste management in Nigeria', *Waste Management*, 29(10), 2787–8.

Subramanian, P. M. (2000), 'Plastics recycling and waste management in the US', *Resources, Conservation and Recycling*, 28(3), 253–63.

Takaoka, Masaki, Nobuo Takeda, Naruo Yamagata and Takahiro Masuda (2011), 'Current status of waste to power generation in Japan and resulting reduction of carbon dioxide emissions', *Journal of Material Cycles and Waste Management*, 13(3), 198–205.

Tanaka, Masaru (1999), 'Recent trends in recycling activities and waste management in Japan', *Journal of Material Cycles and Waste Management*, 1(1), 10–16.

Themelis, Nickolas J. and Scott M. Kaufman (2004), 'State of garbage in America – data and methodology assessment', *BioCycle*, 45(4), 22–6.

Thomas, Christine (1999), 'Waste management and recycling in Romania: A case study of technology transfer in an economy in transition', *Technovation*, 19(6), 365–71.

Troschinetz, Alexis M. and James R. Mihelcic (2009), 'Sustainable recycling of municipal solid waste in developing countries', *Waste Management*, 29(2), 915–23.

Unnikrishnan, Seema and Anju Singh (2010), 'Energy recovery in solid waste management through CDM in India and other countries', *Resources, Conservation and Recycling*, 54(10), 630–40.

Uwadiegwu, B. O. and K. E. Chukwu (2013), 'Strategies for effective urban solid waste management in Nigeria', *European Scientific Journal*, 9(8), 296–308.

Van Haaren, Rob, Nickolas Themelis and Nora Goldstein (2010), 'The state of garbage in America', *BioCycle*, 51(10), 16–23.

Yao, Takeshi, Aretha Aprilia, Tetsuo Tezuka and Gert Spaargaren (2011), 'Municipal solid waste management with citizen participation: An alternative solution to waste problems in Jakarta, Indonesia', in *Zero-Carbon Energy Kyoto 2010*, Springer, Japan, 56–62.

Yusuf, R. O., A. O. Durojaiye and S. E. Agarry (2008), 'Designing an integrated framework for solid waste management in Nigeria', *Journal of Resources, Energy and Development*, 5(1), 1–9.

Zhang, Dong Qing, Soon Keat Tan and Richard M. Gersberg (2010), 'Municipal solid waste management in China: Status, problems and challenges', *Journal of Environmental Management*, 91(8), 1623–33.

Zurbrügg, Christian, Margareth Gfrerer, Henki Ashadi, Werner Brenner and David Küper (2012), 'Determinants of sustainability in solid waste management–the Gianyar waste recovery project in Indonesia', *Waste Management*, 32(11), 2126–33.

5
WASTE PREVENTION ACTION NETS

Hervé Corvellec and Barbara Czarniawska

Introduction

According to the European waste directive (The European Parliament and the Council of the European Union 2008/98/EC), waste policy in the Member States of the European Union is to be organized according to a waste hierarchy. This model ranks waste-management options from best to worst, with waste prevention being the best possible option, followed by re-use, recycling, incineration with energy recovery, and landfilling (Article 4.1).

The rationale for the waste hierarchy includes, among other things, the goal of making consumption more sustainable by reducing the use of resources (Preamble 6), supporting the use of recyclates (Preamble 29), and reducing greenhouse gas emissions originating from waste disposal on landfills (Preamble 3). The Commission stated that 'Waste prevention is closely linked with improving manufacturing methods and influencing consumers to demand greener products and less packaging' (European Commission 2014). More generally, the purpose of the waste hierarchy is to prompt new forms of engagement with waste that reorganize material flows at the precommodity and postcommodity phases of production (Corvellec and Hultman 2012).

The Commission defines waste prevention as:

> measures taken before a substance, material or product has become waste, that reduce: (a) the quantity of waste, including through the re-use of products or the extension of the life span of products; (b) the adverse impacts of the generated waste on the environment and human health; or (c) the content of harmful substances in materials and products.
>
> *(European Commission 2008/98/EC)*

This definition brings two major aspects of waste prevention to the fore: prevention of waste generation and prevention of harm through waste (Arcadis Belgium 2010).

Specific to consumption are eco-labels, awareness campaigns, the development of incentives for clean purchases, and the promotion of re-use and/or repair.

Despite the definitional efforts of European Union authorities, the contours of waste prevention remain blurred. In practice, waste prevention is a broad endeavor that can refer to any of the lifecycle phases of a product or service: design, extraction, production, distribution, use, waste, and end-of-waste (Arcadis Belgium 2010). The best waste-prevention initiatives identified by the European Pre-Waste research project differ widely; they range from the optimizing of packaging for organic food products, to the re-use of furniture, the promotion of decentralized composting, an eco-taxation on disposable plastic bags, the introduction of washable diapers in the nursery, the development of water dispensers, and information about municipal services (Pre Waste 2010).

Many definitions of waste prevention remain debatable. Composting is part of the definition in some countries but not in others. Waste prevention includes re-use because it is performed on non-waste; the preparation for re-use is not considered prevention, however, as it is performed on waste – although it is often next to impossible to distinguish between the two (Arcadis Belgium 2010). It is not possible, therefore, to define waste prevention once and for all, not least because definitions of waste are fluctuating and contextual, despite the European Union's harmonizing attempts. The absence of a clear definition is probably beneficial, considering that many innovative solutions may be yet to come. But even an advocate of diversity must not stop exploring the rationale behind certain waste-prevention initiatives.

The exploration is performed in this chapter with the help of an action net perspective, where action nets are understood as assemblages of collective actions, connected to one another because they are perceived, within a given institutional order, as requiring each other; or, if new, because they are perceived as effective means of accomplishing a goal that lies outside the present order (Czarniawska 2004).

The choice of this perspective is motivated by the goal of moving beyond the ABC (attitude, behavior, choice) model of social change toward sustainability, which dominates the present understanding of social change for environmental transitions and sustainability (Shove 2010). Rather, we demonstrate that waste prevention involves the connection, re-connection, and disconnection of various collective actions, either according to patterns dictated by a given institutional order or in an innovative way (Czarniawska 2010). Waste prevention requires and encourages the construction of new action nets, and/or the reconstruction of existing ones. Effectively connected action nets may stabilize into networks or formal organizations; others may prove temporary. In this way, waste-prevention action nets both build and disrupt the existing institutional order of consumption.

The three empirical examples have been selected to illustrate our claims and the variety of practices in the field: a waste-management company selling waste-prevention services to its waste-producing customers, the opportunity for Swedish households to opt out of unaddressed promotional material, and a car-sharing program. The first example focuses on waste that relates to the consumption of producers, a relatively neglected topic; the second curtails an existing action net rather than constructing a new one; and the third is an example of an initiative that is spreading globally. Before beginning our analysis, however, a closer look at the concept of action nets is necessary.

An action net perspective

The concept of action nets (Czarniawska 2004) originates in a combination of new institutional theory (Powell and DiMaggio 1991) and actor-network theory (Latour 2005). From new institutional theory, it borrows the insight that it is possible in every time and place to speak of a prevailing institutional order, in the sense of an arrangement of institutions that dictates which actions, by convention, should be tied together. In the current institutional order, for example, producers are supposed to try to sell their products, and people with money are supposed to save or invest it. From actor-network theory, the concept of action nets borrows the idea that connecting actions into nets requires the translation of different actions into others, and stabilizing requires the work not only of people, but also of objects. For example, waste collection in residential areas in Sweden presupposes that residents take their containers to the curbside, that waste-collection companies provide adequate vehicles, and that they follow announced collection routes and schedules.

The chronology of an action net perspective is the opposite of the chronology assumed by a conventional network perspective. From an action net perspective, the analysis begins earlier than actor-network theory would suggest, and decisively earlier than the mainstream network theory suggests. According to traditional network theory, actors come first, networks come second, and actions in the network come third. From the action net perspective, actions come first; networks come second (this is where actor-network theory comes in; Latour 2005); and actors, in the sense of such established and recognized units as formal organizations and associations come third. Rather than speaking of actors, actor-network scholars have noted, it is therefore better to speak of "actants" – that which accomplishes or undertakes an act (Greimas and Courtés 1982).

Actants can be individual humans or collectives; they can be artifacts created by humans, such as a machine or a protocol; or they can be natural things such as molecules or animals. This choice of words emphasizes a shift in attention from established actors – who are, in fact, networks – to their origins as humble actants. It caters to the fact that not all actors are constituted as such at all points in the organizing process. It is through the actions they perform successfully that actants become actors; otherwise they remain objects of someone else's actions. It is also from their actions that actors derive an identity. Nobody is born a waste-service provider; but anybody can become one by performing the types of activities associated with a waste service. What matters is the proven ability to act that way.

The term "net" provides a signal that the focus is on connections among actions rather than actors. This focus does not deny the existence of networks of actors, of course; there are a great many actors, from private cliques to large corporations. The point of an action net perspective is to capture organizing at an earlier stage, when things still need to be done, long before powerful actor-networks present themselves to an admiring audience. Actions in action nets are like threads woven or knotted together. If successfully stabilized, they hold in ways that resist tractions and pressure to forces of deformation and displacement.

The action net perspective belongs to processual approaches that focus on organizing (see, e.g., Hernes 2008), designed in contradistinction to essentializing approaches that focus on organizations. The action net perspective targets "what is being done, and how this is connected to other things that are being done in the same context" (Czarniawska 2004). The purpose is to track the process of organizing within organizations and across organizational borders.

The nature of connections between actions is as varied as the human imagination, but it always consists of *translating* the conditions of one collective action into those of another. It can be a matter of mutual adjustment. Recycling centers may hold extended open hours during the Easter weekend, for instance, to accommodate the fact that many individual home-owners use that weekend to clean their gardens for spring. Or the connection can rest on the introduction of a new artifact, as when waste-management companies ask householders to sort food waste in special paper bags or when refill fountains are installed in shops so people can purchase detergent by filling their own containers. Connections can be established by individual human action, as when charities collect second-hand items door-to-door; or they can be mediated by long chains of actants, human or non-human. Such connections must be maintained and, in the case of innovative action nets, perhaps even defended against institutionalized options.

The connecting of actions also requires that actants involve themselves in sensemaking (Weick 1995, 2011) of each other's actions, attempting translations (e.g. Callon 1986; Latour 2005). Translations can thus be understood literally, as talking together and explaining intentions to one another, but also non-linguistically, in the sense of transforming one action into another at the connection point. A great deal of linguistic translation is clearly involved: from one type of specialist vocabulary to another and from one language to another. But perhaps the main point is the translating of actions into one another – by coordination, for example containers are brought to the curb on the day the collection vehicles arrive.

Once the connections between actions have been made and the entire action net is in place, this connection must be stabilized and maintained in good shape (Lindberg and Czarniawska 2006). When relationships among actions are not only stabilized, but also a normative and cognitive fixity (that is, they can be justified in an appropriate vocabulary and taken for granted), they will become the basis for actors to acquire character ("He is a pioneer of waste prevention") and allow them to build networks ("No need to change those providers; we can rely on them").

Not all connections between actions will become stable, however, and a researcher's interest in an action net lies in showing whether or not and how ongoing processes of organizing practices build stable relationships (Lindberg and Walter 2013). Another aspect of the construction of new action nets is the extent to which such innovative nets draw upon, adapt to, or change the existing institutional order.

Three examples of waste prevention

In what follows, the action net perspective has been used to analyze three cases of waste prevention in Sweden. The first case is the only example of corporate waste discussed here; it focuses on the production stage – a waste-management company selling waste-prevention services to its waste-producing customers (NSR, Helsingborg). The second case concerns household waste: an attempt at waste prevention by providing the opportunity to opt out of unaddressed promotional material ("No advertising, thank you" signs) offered to Swedish householders. The third case is a car-sharing program (Sunfleet, Sweden) that illustrates the construction of an alternative pattern of consumption. The concept is globally spread, but we focus on a local example.

Waste-reduction services

NSR is an advanced solid-waste management company co-owned by six municipalities in the Northwest Scania region of Southern Sweden, and as such it is responsible for waste-collection and waste-treatment services in the entire region. NSR is one of the major biogas producers in Sweden; a producer of biofertilizers; and, more generally, a company with competence in biological treatment, waste characterization, recycling, and landfill research (NSR 2013).

Since 2007, NSR has offered tailored waste-reduction and waste-prevention services to waste-producing companies. This offer required the construction of a new action net, which in turn required the creation of incentives. Larger companies or companies with environmental goals were offered a comprehensive waste-management contract, with personal contact, proximity to services, and an overview of the waste-management situation (NSR 2011) – thus, an entrance to a network. Smaller companies were offered effective management of waste streams, with custom waste collection and proximity to efficient service (NSR 2011) – thus an invitation to join the action net. Moreover, NSR provides hazardous-waste consulting services to businesses throughout the region, offering the services of its chemists and safety advisers in the classification and handling of all categories of hazardous waste, with the exception of radioactive waste (NSR 2007). The offer of waste-prevention services is part of the mission given to NSR by its owners (Helsingborg Stad – Kommunfullmäktige 2012), part of its business model as a waste-management company (Corvellec and Bramryd 2012; Corvellec *et al.* 2012), and a way for the company to innovate in order to climb the waste hierarchy (Corvellec *et al.* 2013) and contribute to the sustainability of consumption.

Several action nets had to be initiated and coordinated earlier to give life to these prevention services. NSR had to canvass waste producers within its geographical zone of activities; proceed to systematic and standardized waste analyses to assess the kind and quantity of waste delivered by waste producers with an interest in these services; design custom-made waste-management solutions for the materials in this waste, identify how to process and where to sell them; collaborate with the waste producer to redesign its internal material management processes to reorient material flows from waste to recycling; and introduce economic incentives for waste producers to enter a waste-reduction program, while maintaining its own profitability.

To connect their actions to those of NSR, waste producers must integrate NSR's view into their material processes designs – translate those designs so that they fit this view. They need to redesign their work processes to replace non-recyclable material with material that NSR can recycle; install dedicated waste and material containers to sort their key waste streams by volume, worth, and toxicity; and introduce incentive schemes to promote and monitor the internal waste-prevention policy – sharing the savings with employees, for example. Many of these actions have been undertaken in common by representatives of each company; but many other actions have been taken by NSR and waste producers with their own suppliers and customers.

NSR and vegetable wholesalers have developed a separate collection and processing system for unsold vegetables. Instead of being mixed with other waste, unsold vegetables are pressed; the water they contain is drained off into wastewater, nutrients are collected in a form pure enough to be fed directly into NSR's biogas production chain, and only the packaging becomes waste. This process reduces costs for wholesalers by reorienting their waste toward the wastewater-management system and increases their income by connecting the remaining material to energy production; the new action nets are acting on both ends of the economic value creation process.

In the case of a local thermal insulation manufacturer, NSR analysis led to ways of reprocessing by-products to turn them into a filling material that can be sold, rather than them ending up as an inert material in an expensive landfill. It is noteworthy that this waste-prevention action net played a key role in the decision to maintain the thermal insulation plant when the international head office had decided to terminate one of three similar plants in other European locations. Establishing a new action net prevented this actor from being reduced to an actant with undetermined identity and an uncertain future.

A press for unsold vegetables and a machine to upgrade waste into filling material are both highly visible stabilizers of the connections that have been built to establish waste-prevention services. But no less important are the less visible connections upon which NSR's waste-prevention action net depends: common definitions of waste and materials; agreements on collection frequencies; and, more generally, a shared view of the relevance of thinking in terms of waste prevention. There is more than the visible to action nets.

"No advertising, thank you" signs

The second case involves an action net that grafted itself onto an existing action net in order to reduce the latter. In 1993, the Swedish Royal Postal Agency (now called Posten AB), the Swedish Consumer Agency, and the Association of Swedish Direct Marketing Companies (SWEDMA) agreed to work together to address the issue of direct marketing. This agreement provided Swedish householders with the opportunity to opt out of unaddressed promotional material from door-to-door mail distributors by simply placing a "No advertising, thank you" sign beside their mailbox or letter slot. Because the original sign did not stop civic information such as bus timetables, information from political parties, or free newspapers from being distributed, householders were also given the opportunity to post a "No free newspaper" sign by their mail slot. They can also refuse addressed advertising by listing themselves at a central register. Likewise, private individuals can register centrally to indicate that they do not want direct marketers to telephone them (Konsumentverket 2013; Svensk Direktreklam n.d.).

The limiting of advertising has been an established policy to reduce the consumption of such products as alcohol, tobacco, and prescription medicines, even if the impact of advertising on aggregated demand may not be significant for either drinking (Wilcox *et al.* 2012) or smoking (Capella *et al.* 2008), and people may be neutral on the appropriateness of advertising medicine (Miller and Waller 2004). A public ban on advertising can also be a means of protecting specific groups such as children from obesity (Dhar and Baylis 2011). In this case, however, it was not a matter of a public ban, but of a multipartite agreement to offer an opt-out option to householders. Householders can make the deliberate choice of placing a "No advertising, thank you" sign on their doors, possibly shifting the blame for excessive consumption on advertisers and marketers (Pereira *et* al. 2012).

For householders to be able to stop unaddressed advertisements requires scores of actions to be coordinated into an action net. The efficacy of the *No ads* or *No free newspaper* signs depends on SWEDMA reminding its member organizations of the necessity of respecting these signs, and the Swedish Consumer Agency's handling of complaints about failure to respect them. Registers for people to record their wish not to be disturbed by unsolicited phone calls must be connected to the databases that Swedish telemarketers use, and householders must monitor the calls by reminding telemarketers that they are actually not allowed to phone them. A continual connecting and reconnecting of these actions is necessary for the goals of the scheme to be achieved.

One noteworthy aspect of the *No-ads* scheme is the argument of some commercial actors: Opting out may cause people to miss crucial information – when a car is due for its mandatory annual checkup, for example (TV4 2013). Clearly, the *No-ads* action net runs against interests that find their expression in direct-marketing action nets. The purpose of the *No-ads* action net develops in competition with these nets and, more generally, conventional commercial action nets, to limit (some people would say "damage") their reach. The two compete in their attempts to impact consumers' behavior.

A car-sharing program

A product of yet another waste-reduction action net is Sunfleet, a business-to-consumer car-sharing service. The company was started in the early 1990s by

Hertz and Volvo, as a way of filling the market segment between permanent car ownership and occasional car rental. The service was developed around the notions of convenience, flexibility, cost effectiveness, and sustainability. Cost effectiveness here refers to an absence of fixed costs and the opportunity for individuals to monitor their car transportation costs. Sustainability refers to the possibility of choosing the size of car that exactly fits the need of the moment, to the incentive to choose other means of transportation such as cycling or public transportation whenever possible, and to Sunfleet's choice of fuel-efficient vehicles, often less than two years old. To clarify the contribution of car sharing to sustainability, the company quotes the Swedish Transport Administration's claim that one car in a sharing program replaces five individually owned cars. The company's commercial motto is "A car only when you want it". Sunfleet presents itself as a "car revolutionary," claiming to lead, together with its members, the way toward a more sustainable mobile society. It declares that its goal is to introduce car sharing in Sweden – to change how Swedes look at owning and using a car (Sunfleet 2013).

Car sharing is not only emblematic of an innovation in the logic of service (Michel *et al.* 2008). It is also emblematic of a consumption based on access (Bardhi and Eckhardt 2012) and, more generally, of sharing as an alternative to possession (Belk 2010). Car sharing is also emblematic of an evolution of mobility consumption toward greater sustainability (Antonio *et al.* 2012).

To develop a car-sharing action net, Sunfleet had to produce a significant change in the relationship between the car user and the car (Michel *et al.* 2008). It had to coordinate its own actions and develop new types of connections among the actions of car producers, financing bodies, and car-maintenance companies, but also with the municipality and other land owners that provide parking lots, and, not least, with people interested in trying this type of service who need to disconnect the idea of mobility and the freedom attached to it from the idea of owning a car. In addition, Sunfleet has established collaborations with housing companies and such organizations as the City of Gothenburg and the University of Gothenburg to offer packaged solutions for carless urban dwellers. Part of these solutions is the hyperlinks that direct Sunfleet members toward environmental education programs, supporting a ride-sharing community, and hiring electric bicycles – all literal expressions of the connective logic of action nets.

Fitting the local institutional context is helpful, of course (Mont 2004). But the car-sharing action net rests primarily on well-functioning economic, legal, technical, and behavioral connections among the actions described here. And these connections need to be maintained and redesigned whenever any of the actants and actors change their ways of doing things – that is to say, all the time. Such ruptures in the connections as poorly maintained vehicles or an ill-functioning booking system would dissolve the action net and effectively stop the service. Trust, commitment, respect for such rules as punctuality and, more generally, the commons are implicit actants of car-sharing systems. Along with the right incentives (Lerner 2012), service innovations depend on reliability for their success.

Action nets for waste prevention

Several insights can be gained from analyzing these three waste-prevention initiatives from an action net perspective.

First, it is possible to speak of waste prevention because the connections between the actions in the nets have reached some level of stability, at least temporarily. A sign of this stability is the fact that the nets are no longer dependent on the idiosyncrasies of actants. The interactions are stable enough that a waste producer, a car manufacturer, or a direct marketing company can replace one another. They become stabilized to the point at which they can be seen as a pattern to be imitated – after all, Sunfleet is not the first car-sharing company in the world. The *No-ads* scheme has served as a source of inspiration to establish a method for saying "no" to unsolicited advertisement in mobile phone ads. Stabilization does not mean lack of change, however. Action nets are constantly evolving. Actants can agree to change some aspects of the *modus operandi* of their interactions – to answer to changes in legislation or let the activity evolve, for example. Entrepreneurial actants and actor-networks can include new actions into existing nets, or even connect action nets to one another. Nets can develop in unexpected directions. Some companies have even started selling ready-to-use *No ads* signs. Likewise, texts like this chapter can connect these initiatives to new actions if they are considered a source of inspiration by waste decision-makers. Stability does not mean that actions within the net remain the same.

A *second* noteworthy aspect of these three waste-prevention initiatives is the shape they give to specific perceptions of waste prevention: reducing the volume of unaddressed advertisements being produced and distributed; integrating the constraints and possibilities of contemporary waste management in the design of material management processes; and reorienting people from ownership to rental and use, which is a key tenet of what is called *économie de la fonctionnalité* in French (Bourg and Buclet 2005) – literally the economy of functionality, but unfortunately translated as service economy. Each initiative is an effort to reduce the volume of material throughput (Daly and Farley 2004) in the economy. Furthermore, it decouples waste production from economic growth, which is one of the key goals of globally sustainable waste management (UNEP 2011). These ideas may have existed before the first actions were taken; they may have emerged slowly along with the organizing process; or they may be born only after the process is complete – when people make sense, retrospectively, of what they have done. Action nets materialize visions, but into dynamic processes rather than static structures.

A *third* insight concerns the key role played by artifacts in action nets. Human actants appear to depend on good relationships with their non-human counterparts if they want to connect waste-reducing actions. All three initiatives in this study depended on dedicated technological devices: waste containers, signs, roofed facilities, or booking systems. Action nets must also connect well to such places as householders' letter slots for the *No-ads* scheme or the Helsingborg region for NSR's waste-prevention services. Waste prevention may aim at de-materialization, but artifacts are central to the construction and maintenance of waste-prevention connections.

Fourth, waste-prevention initiatives can also be considered relational spaces. A relational understanding of space (Shields 2013) suggests that spaces are constituted by a series of practices and materials that determine their character. Action nets can travel – literally – but such travels require effective translations that make the nets fit for the place they land (Czarniawska and Sevón 2005). An action net perspective emphasizes the fact that waste-prevention initiatives are global, but only to a point. They are also eminently local, not least due to the localness of the interactions from which they emerge.

Fifth, action nets are always being constructed in relation to existing action nets. On the one hand, new action nets build on existing nets. NSR's waste-prevention services build on existing nets that allow recycled materials to re-integrate production processes in particular and the economy in general. The *No-ads* initiative builds on an existing collaboration between public authorities in charge of consumption and the direct marketing industry. Sunfleet relies on the existing automobility infrastructure.

On the other hand, new action nets challenge existing nets. NSR's waste-prevention services challenge existing waste-handling action nets that thrive on increasing waste volumes. The *No-ads* initiative is an effort to limit the spread of the direct-marketing action net. Sunfleet's car-sharing service creates an alternative to car ownership and car rental nets. New action nets coexist with existing action nets in many ways, sometimes coexisting, sometimes complementing one another, sometimes competing.

The challenges of waste prevention

Our analysis has demonstrated that waste prevention requires the invention, development, and stabilization of new connections between collective actions. Translating the goals expressed in the highest step of the European waste hierarchy into concrete results requires the invention, development, connection, and stabilization of nets of actions based on new and often innovative understanding, priorities, habits, and artifacts.

An action net perspective focuses on the dynamics of organizing. It clarifies the fact that waste prevention is a matter of developing new connections among collective actions that will hopefully result in less, if any, waste, according to the European Union's definition of waste, and less adverse impact on the environment. It also underscores the need to understand transition policies as the creation of new infrastructures that make possible but also constrain the emergence of new action nets.

Action nets are constrained by the existing institutional order, but they challenge this order as well. Our analysis indicates that waste prevention is disruptive of the institutional order of consumption. Initiatives based on the prevention of waste tend to aim at slowing or reducing material flows in society. As such, they have the potential to attract the ire of many types of people – advocates of the growth paradigm upon which rest many current business welfare models in contemporary societies, for example. One can therefore expect protests by the actors who are against the introduction of waste-prevention patterns of production, distribution, consumption, and disposal.

Our analysis also indicates that the "higher step of the waste hierarchy" is not above the lower steps. It stands for a rupture. Unlike the initiatives at the lower steps of the waste hierarchy (e.g. incineration or recycling), in which waste is considered as a resource to exploit, waste prevention challenges the existing order of wasting. Prevention runs against the traditional reasoning, which merely addresses *existing* waste. Whether it is a matter of preventing increases in energy use, urban transportation, or greenhouse gases, the rationale of prevention differs in character from the rationale of managing that which already exists. Management assumes a normalization of waste (Corvellec 2014), whereas prevention involves preventing something coming into being. Non-existence, rather than diligence, is the measure of the performance of prevention, which requires innovative action nets.

Indeed, the initiatives discussed here suggest that a new institutional order of waste may be under development, as other studies have implied. In this new institutional order, "wasting less" could become the leading societal narrative (Corvellec and Hultman 2012). Waste-management companies would develop business models based on circularity and waste prevention; excessive consumption (Urry 2010) would be questioned; and shopping for second-hand items would become a cultural standard (Gregson and Crewe 2003). In such an order, waste governance is pluralistic (Zapata Campos and Zapata 2013) and multi-leveled (Bulkeley et al. 2005), and global waste prevention is a necessary part of sustainable urban development (Zapata and Hall 2013).

Finally, an action net perspective on waste prevention encourages a key question: Why do not all efforts at connection become stable? Actor-network theory (Latour 2005) states that projects fail because some participants do not manage to translate the interests of other participants and therefore fail to align them. Action net theory confirms this observation, but adds that actions are sometimes not translatable into one another, or the connection points are not maintained with enough care. With all interests aligned, actants can see the net they connected dissolve – because they did not take enough care, because the stabilizing artifacts were not resilient enough, or because the existing networks destroyed it. Such a negative case should provide a fascinating study, even though actor-networks tend to hide failures, making them difficult to document.

The analysis of waste prevention made from the perspective of action nets demonstrates that, in practice, the diversity of waste-prevention initiatives is doubled by a diversity of actions specific to each initiative. The richness and intricacy of connecting diverse actions and maintaining these connections as conditions change strongly contrasts with the linear simplicity of the European waste-hierarchy model. We believe that developing waste prevention would benefit from a recognition of the far-reaching diversity of waste-prevention actions – a critical recognition, not least for the construction of waste-governance models for waste prevention. If waste prevention is to improve the sustainability of consumption, it is not by step by step, but connection by connection, action by action.

Acknowledgments

We are grateful to María José Zapata Campos and Richard Ek for their comments on previous versions of this text. We are also grateful to Nina Colwill for her editorial help. Hervé Corvellec also acknowledges the kind hospitality of the Department for Organization at the Copenhagen Business School (Denmark), where he wrote parts of this text. This study is part of the research project *From waste management to waste prevention – Closing implementation gaps through sustainable action-nets* financed by the Swedish Research Council FORMAS (Environment, Agricultural Sciences and Spatial Planning).

References

Antonio, Musso, Corazza Maria Vittoria, and Tozzi Michele (2012), "Car sharing in Rome: A case study to support sustainable mobility," *Procedia – Social and Behavioral Sciences*, 48, 3482–91.
Arcadis Belgium (2010), *Final Report: Analysis of the Evolution of Waste Reduction and the Scope of Waste Prevention, European Commission Dg Environment*. Antwerpen: Arcadis.
Bardhi, Fleura and Giana M. Eckhardt (2012), "Access-based consumption: The case of car sharing," *The Journal of Consumer Research*, 39, 881–98.
Belk, Russell (2010), "Sharing," *Journal of Consumer Research*, 36(5), 715–34.
Bourg, Dominique and Nicolas Buclet (2005), "L'économie de fonctionnalité: Changer la consommation dans le sens du développement durable," *Futuribles*, 313, 27–38.
Bulkeley, Harriet, Matt Watson, Ray Hudson, and Paul Weaver (2005), "Governing municipal waste: Towards a new analytical framework," *Journal of Environmental Policy and Planning*, 7(1), 1–23.
Callon, Michel (1986), "Some elements of a sociology of translation: Domestication of the scallops and the fishermen of St Brieuc Bay," *Sociological Review Monograph*, 32, 196–233.
Capella, Michael L., Charles R. Taylor, and Cynthia Webster (2008), "The effect of cigarette advertising bans on consumption," *Journal of Advertising*, 37(2), 7–18.
Corvellec, Hervé (2014), "Recycling food waste into biogas, or how management transforms overflows into flows," in Barbara Czarniawska and Orvar Löfgren (eds), *Coping with Excess: How Organizations, Communities and Individuals Manage Overflows*, Cheltenham: Edward Elgar, 154–72.
Corvellec, Hervé and Johan Hultman (2012), "From 'less landfilling' to 'wasting less': societal narratives, socio-materiality and organizations," *Journal of Organizational Change Management*, 25(2), 297–314.
Corvellec, Hervé and Torleif Bramryd (2012), "The multiple market-exposure of waste management companies: A case study of two Swedish municipally owned companies," *Waste Management*, 32(9), 1722–7.
Corvellec, Hervé, Torleif Bramryd, and Johan Hultman (2012), "The business model of solid waste management in Sweden: A case study of two municipally-owned companies," *Waste Management & Research*, 30(5), 512–18.

Corvellec, Hervé, María José Zapata Campos, and Patrik Zapata (2013), "Infrastructures, lock-in, and sustainable urban development: The case of waste incineration in a Swedish metropolitan area," *Journal of Cleaner Production*, 50(1), 32–39.

Czarniawska, Barbara (2004), "On time, space, and action nets," *Organization*, 11(6), 773–91.

— (2010), "Going back to go forward: On studying organizing in action nets," in Tor Hernes and Sally Maitlis (eds), *Process, Sensemaking, and Organizing*, Oxford: Oxford University Press, 140–60.

Czarniawska, Barbara and Guje Sevón (2005), *Global Ideas: How Ideas, Objects and Practices Travel in the Global Economy*. Malmö: Liber.

Daly, Herman E. and Joshua C. Farley (2004), *Ecological Economics: Principles and Applications*, Washington (DC): Island Press.

Dhar, Tirtha and Kathy Baylis (2011), "Fast food consumption and the ban on advertising targeting children: The Quebec experience," *Journal of Marketing Research*, 48(5), 799–813.

European Commission (2014), "Waste." Available online at http://ec.europa.eu/environment/waste/ (accessed 27 August 2014).

The European Parliament and the Council of the European Union (2008/98/EC), "Directive 2008/98/Ec of the European Parliament and the Council on Waste and Repealing Certain Documents," *Official Journal of the European Union L 312/3*, 0003 - 30.

Gregson, Nicky and Louise Crewe (2003), *Second-Hand Cultures*. Oxford: Berg.

Greimas, Algirdas Julien and Joseph Courtés (1982), *Semiotics and Language: An Analytical Dictionary*. Bloomington: Indiana University Press.

Helsingborg Stad - Kommunfullmäktige (2012), *Ägardirektiv För NSR (Nordvästra Skånes Renhållnings Ab) [Owners' Directive]*, Helsingborg: Helsingbiorg Stad.

Hernes, Tor (2008), *Understanding Organization as Process: Theory for a Tangled World*. Abingdon: Routledge.

Konsumentverket (2013), 'Nej till reklam'. Available online at www.konsumentverket.se/Lagar--regler/marknadsforing-och-reklam/Nej-till-reklam/ (accessed 29 October 2013).

Latour, Bruno (2005), *Reassembling the Social: An Introduction to Actor-Network-Theory*. Oxford: Oxford University Press.

Lerner, Joshua (2012), *The Architecture of Innovation: The Economics of Creative Organizations*. Oxford: Oxford University Press.

Lindberg, Kajsa and Barbara Czarniawska (2006), "Knotting the action net, or organizing between organizations," *Scandinavian Journal of Management*, 22(4), 292–306.

Lindberg, Kajsa and Lars Walter (2013), "Objects-in-use and organizing in action nets: A case of an infusion pump," *Journal of Management Inquiry*, 22(2), 212–27.

Michel, Stefan, Stephen W. Brown, and Andrew S. Gallan (2008), "Service-logic innovations: How to innovate customers, not products," *California Management Review*, 50(3), 49–65.

Miller, Kenneth E. and David S. Waller (2004), "Attitudes towards DTC advertising in Australia: An exploratory study," *International Journal of Advertising*, 23(3), 389–405.

Mont, O. (2004), "Institutionalisation of sustainable consumption patterns based on shared use," *Ecological Economics*, 50(1–2), 135–53.

NSR (2007), *Putting the Environment First*. Helsingborg: Nordvästra Skånes Renhållnings AB.

— (2011), "Helkund Hos NSR." Available online at http://www.nsr.se/Default.aspx?ID=38 (accessed 29 october 2013).

— (2013), *Årsredovisning 2012 [Annual Report 2012]*, Helsingborg: Nordvästra Skånes Renhållnings AB.

Pereira Heath, M. T. and A. Chatzidakis (2012), "'Blame it on marketing': Consumers' views on unsustainable consumption," *International Journal of Consumer Studies*, 36(6), 656–67.

Powell, Walter W. and Paul J. DiMaggio (1991), *The New Institutionalism in Organizational Analysis*. Chicago: University of Chicago Press.

Pre Waste (2010), *Mapping Report on Waste Prevention Practices in Territories within EU27*. Brussels: European Union.

Shields, Rob (2013), *Spatial Questions: Cultural Topologies and Social Spatialisation*. London: Sage.

Shove, Elizabeth (2010), "Beyond the ABC: Climate change policy and theories of social change," *Environment and Planning A*, 42(6), 1273–85.

Sunfleet (2013), 'Om Sunfleet'. Available online at www.sunfleet.com (accessed 29 October 2013).

Svensk Direktreklam (n.d.), 'Nej tack till reklam'. Available online at http://www.delareklam.nu/index.php?page=om-oss (accessed 29 October 2013).

TV4 (2013), 'Nej till reklam kan leda till körförbud'. Available online at http://www-origin.tv4play.se/program/nyheterna-%C3%B6stersund?video_id=2307445 (accessed 29 October 2013).

UNEP (2011), *Decoupling Natural Resource Use and Environmental Impacts from Economic Growth, A Report of the Working Group on Decoupling to the International Resource Panel*. Nairobi, United Nations Environment Programme.

Urry, John (2010), "Consuming the planet to excess," *Theory, Culture & Society*, 27(2–3), 191–212.

Weick, Karl E. (1995), *Sensemaking in Organizations*. Thousand Oaks, CA: Sage.

—— (2011), "Reflections: Change agents as change poets – on reconnecting flux and hunches," *Journal of Change Management*, 11(1), 7–20.

Wilcox, Gary B., Kim KyunkOk Kacy, and Heather M. Schulz (2012), "Liquor advertising and consumption in the United States: 1971–2008," *International Journal of Advertising*, 31(4), 819–34.

Zapata Campos, María José and Patrik Zapata (2013), "Switching Managua On! Connecting informal settlements to the formal city through household waste collection," *Environment & Urbanization*, 25(1), 1–18.

Zapata, María José and Colin Michael Hall (2013), *Organising Waste in the City: International Perspectives on Narratives and Practices*. Bristol: The Policy Press.

6
CURBSIDE CARTOGRAPHIES IN AN URBAN FOOD-WASTE COMPOSTING PROGRAM

John W. Schouten, Diane M. Martin and Jack S. Tillotson

Introduction

This chapter draws from a larger research project wherein we take an actor-network theory (ANT) approach to examining the creation, implementation and aftermath of a program of residential curbside collection of food waste for composting in the city of Portland, Oregon, USA. Our ultimate aim is to better understand the kinds of actors, assemblages and translations with real promise for making contemporary systems of production and consumption more environmentally sustainable and socially equitable (Martin and Schouten 2012). Social science approaches to problems of consumerism and sustainability have repeatedly run up against an intransigent attitude–behavior gap (Kilbourne and Pickett 2008; Kollmuss and Agyeman 2002). We believe the problem may lie in prioritizing consumer agency and the idealist notions that cultural change begins with cognition, values and attitudes. We begin by discussing actor-network theorization and its appropriateness for this study. We then describe our methodology. Our findings report the apparent success of the municipal food scraps composting program (hereafter referred to simply as "the program") and the key elements of that success. We discuss actors, both human and non-human, that contributed to the overall performance of the program, in some cases threatening to destabilize or undermine it.

ANT in the world of waste

Science and technology scholars (Callon 1986; Latour 1987; Law 1988) introduced ANT to the world as a remedy to perceived shortcomings of social-constructionist approaches, which these proponents claimed give short shrift to the roles of technology and other non-human actors in organizing social life. Actor-network theorists are clear on the point that ANT is not a theory, but rather a constructivist (Latour 2005) ontological epistemology (Law 2004) for understanding and theorizing social phenomena. ANT is one member of a whole family of theories, philosophies and research approaches resurgent in a variety of disciplines with the intent of reintroducing materiality to understandings of the social. Under the family name of New Materialism (Dolphijn and van der Tuin 2012) we find interesting convergence among ANT scholars and others of a materialist bent. ANT is an evolving pursuit with a history of fluidity and reflexivity that makes it difficult to pin down, even for its creators. Latour (2005: 88) writes that, "ANT is the story of an experiment so carelessly started that it took a quarter century to rectify it and catch up to what its exact meaning was". That exactness has still never been fully realized as ANT continues to evolve. One recent evolution has been its embrace of terminology beginning with Deleuze and Guattari (1988) and propagated by DeLanda (2006). Latour (1999) himself has repented of the term actor-network and admitted affection for Deleuze's metaphor of rhizomes, which are more subtly and completely interpenetrated than is generally assumed for relationships in a network. Callon has begun using terms such as agencement, arrangement or assemblage (Çalışkan and Callon 2010) as opposed to actor-networks, bringing him closer in line to the thinking of DeLanda. Palmås (2007) points out that such use of Deleuzian language by actor-network theorists doesn't indicate a complete conversion to Deleuzian thought. ANT remains ANT even though some interesting translations have been occurring within it. For now, it seems perfectly permissible to use the terms actor-network and assemblage more or less interchangeably, which we do.

We take an ANT position knowing that in anthropology there is also a well-respected tradition of theorizing the roles of material objects and infrastructures in shaping and maintaining social relations (Appadurai 1986; Gell 1992; Harris 1979; Miller 1998). Appadurai gets credit for launching much of the recent anthropological thinking about the roles of materiality in organizing the social (Van Binsbergen 2005). In his introduction and edited volume on *The Social Life of Things*, Appadurai (1986) focuses on the means whereby material objects become commodities with economic value. This work promises to shed light on how consumption coupled with production systems can be more environmentally sustainable and socially equitable. That is, the composting of household food waste transforms something with less than no value (being a nuisance or a burden) into something with value and even marketability for gardeners and farmers. For enthusiasts of organics and "green" solutions, the effects of the conversion of food waste to compost may even verge on enchantment (Gell 1992).

That material actors, such as natural topographies or technological infrastructures, cause and limit human response is an ANT tenet. The position of ANT on human agency is that it exists in analytic symmetry with the agency of non-human actors and that change occurs as a product of the relations between them. We recur to ANT to examine the co-constitutive relations among policy makers, consumers, waste, the infrastructure for its collection and transformation, and the discourses surrounding it. The core construct in ANT is the actor-network, which is understood as a heterogeneous assemblage of human, non-human and hybrid actors. The assemblage is constantly performed through various kinds of translations of things and meanings. These translations are realized within the relations among the various actors (Callon 1986; Law 2008). The relations are generative and together may work to assemble, stabilize or destabilize a network. A social program, such as the waste management program here under consideration, is clearly a heterogeneous assemblage. Among its human actors are consumers, trash carriers and government officials. Non-human actors include discourses such as regulations, collection schedules and news stories. Other actors are technological, among them being processing facilities, garbage bins, recycling bins, trucks and much of the refuse produced in consumer households. Still other actors are biological or chemical, such as food waste, animal pests and odors from decomposition processes.

Any transformation or movement of materials or meanings from one medium, form or space to another constitutes a translation (Latour 2005). Callon (1986), however, uses the term more specifically to name a process whereby one actor problematizes a situation and then identifies, enlists and mobilizes other actors into a network to deal with it, usually attempting to take and maintain a position as an obligatory passage point through which everything else in the network is monitored and processed. This type of translation can be a tricky business, because resistance or deviance from certain actors can subvert an entire project or bring it to naught (Akrich et al. 2002; Callon 1986). Such is often the case when a firm, having developed a new product, attempts to create a market for it. A market is a complex actor-network (Çalışkan and Callon 2010) and the refusal of any key actors – prospective customers, for example – to participate can turn the firm's efforts into an expensive object lesson (Martin and Schouten 2014). Even in existing assemblages, stability of any sort requires continuous translation (Thomas et al. 2013). The City of Portland's creation and implementation of the curbside food-waste collection program is a classic case of translation as theorized by Callon (1986). Due to the agency of numerous actors, the stability of the resulting assemblage was anything but guaranteed.

ANT is especially appropriate for this study for several reasons. For one, the emphasis on the agency of non-human subjects is useful. We are especially interested in the roles of discourses, technologies and infrastructures as implements of cultural change. A bias toward seeing these actors as agentic subjects heightens awareness of their efficacy in affecting behaviors at the individual and community levels. Second, the perspective of assemblages as on-going translations emphasizes their mutability more than their stability, which we believe may also facilitate understanding of the relational dynamics at work in the program we are studying. Bringing an ANT perspective to the problem may, by giving nature, technology and infrastructure their rightful due as actors, help break through the logjam that is the attitude–behavior gap.

Finally, in a study of market dynamics and emergence Martin and Schouten call attention to a promising potential of ANT, namely that of:

> identifying the conditions in which a network is susceptible to major translations from strategic interventions and determining what kind of catalyst is

likely to effect the desired change. For example, inasmuch as markets are implicated in the problems of unsustainable production and consumption, it would be useful to understand where there might exist leverage points for altering market structures or dynamics.

(Martin and Schouten 2014: 868)

This study takes one step in the direction of realizing that potential in the realm of waste management and reduction.

Research team and context

As the researchers in this study we must reveal that we regard ourselves as environmentalists, relatively well versed in the natural-science arguments around environmental sustainability. As such, we are advocates of a particular definition of and approach to sustainability developed and promoted by the international NGO, The Natural Step (www.thenaturalstep.org), and we have used its strategic framework in our own writings about sustainability and marketing (Martin and Schouten 2012).

Portland, Oregon, is considered to be one of the most sustainable American cities (Thompson 2009) and provides the setting for this research project. The city government features a bureau of sustainability and planning that is responsible for developing and implementing "policies and programs that provide environmental, economic and social benefits to residents, businesses and government" (City of Portland n.d.). This city of about 1.5 million people prides itself on active neighborhood associations, strong environmental protections, and abundant public parks and green spaces. Portland vies for the top spot for bicycle use (*Bicycling* 2012) even though it is equally famous for its rainy character.

The program that is the focus of this study was introduced to all Portland households (excluding apartment blocks), including renters and homeowners, in October 2011 after a pilot program in selected neighborhoods with selected trash haulers. Prior to the program's implementation, trash and non-food recyclables including paper, cardboard, metals and glass were collected on a weekly basis. Every other week, haulers also collected yard trimmings. Food waste was not collected and therefore was relegated to the trash and then the landfill. The few exceptions included the small-scale composting of plant matter by some individual consumers for personal use in gardens and the disposal of food waste through in-sink disposal machines into the wastewater system. With the introduction of the program, households were allowed to put all food waste, including both plant and animal matter, in with their yard trimmings, which were then collected every week rather than every other week. To keep collection costs at close to the same level, and to encourage active participation in the program, trash collection was reduced to once every other week. The result was that haulers made the same number of trips and emptied roughly the same number of containers. Only the contents of the containers shifted.

Methods

Our data to date include in-depth interviews with the two officials of the City of Portland Department of Planning and Sustainability most centrally involved in the program, communication to consumers from the city and from trash haulers regarding the program, and a comprehensive collection of articles and blogs about the program including public responses to them from Portland's primary newspaper, *The Oregonian*.

The data and analysis presented here are part of a larger research effort. In keeping with the onto-epistemological guidelines of ANT, we are constructing an understanding of the program as an assemblage in constant translation, focusing our gaze on the interactions and relations among the key actors, mapping the various translations and exploring the agentic efficacy manifested therein.

Findings

A successful translation

Portland, Oregon has a history of vigorous public participation and civic resistance, especially at the level of neighborhood associations, and these associations have real power to influence public policy (Witt 2004). Despite the environmental nature of the composting program, this tradition of civic resistance creates problems for the implementation of municipally mandated programs. However, from planning through to implementation and subsequent management, the composting program has been a successful example of Callon's translation process.

Problematization

Problematization of the food waste situation began at the City of Portland (the City) and was undertaken specifically by the Department of Planning and Sustainability (DPS). They defined the problem as too much food waste going to landfill sites, which lie about ninety miles away by truck up the Columbia River Gorge. As trash, the food waste contributed significantly to trash volume and hauling costs as well as to methane production through eventual decomposition. The City found an alternative solution in commercial composting whereby the same waste matter could be converted to high-quality fertilizer, thus transforming a costly waste stream into a source of value. Consumers in neighboring Seattle, Washington's voluntary curbside composting program adopted such a program in 2005. Both trash and compost are collected in Seattle every week. The City adopted this idea of separate trash and compost collection, but with an important difference: Portland's trash would be collected every other week while compost would be collected every week.

Interessement

The next stage of the translation process involved identifying the actors that would be necessary to the process of planning and implementation. DPS identified trash haulers, which consisted of over a dozen independent companies that interact directly with the customers on their routes, and which have a stake in routing, pricing and infrastructural decisions; commercial composting companies, which would have to capitalize expanded capacity; and citizens.

Enrollment

Once key actors were identified, they needed to be engaged in the program as participants and problem solvers. This began as series of dialogs among DPS officials and commercial trash haulers, in which topics such as logistics and pricing were resolved. Eventually it culminated in a pilot program involving four neighborhoods with certain residential haulers that volunteered to participate and to enroll along with them the households on their routes.

Mobilization

A successful pilot program in four neighborhoods did much to convince other haulers to participate in the program. Two thousand homes in geographically and socio-economically diverse neighborhoods served as testing grounds (Har 2010). In a deviation from usual governing practice in Portland, DPS sought no citizen input into the program other than the feedback through the agency's website in the pilot program. The pilot curbside composting program showed significant potential in overcoming the landfill problem, and the City rolled out the curbside program for all single-family homes.

Early impacts: behaviors and attitudes

Ecosystem impacts of the program have been substantial and immediate. Oregon Public Broadcasting (Sept. 13, 2012) and Portland DPS report that at the program's one-year mark: "the city has eliminated 40 per cent of its residential garbage .... That translates into cost savings. Food scraps cost the city 43 per cent less per ton to haul away." By that one important metric of behavioral change the program has been successful.

One respondent to an *Oregonian* survey (Slovic 2012b) refers to the role of program as an actor in changing behaviors:

> We had a composting bin in the back yard for five years; however, it was really more of a worm bin than a compost bin. For most of the year, we couldn't maintain the temperatures needed to effectively compost all of our kitchen/yard waste.... I know the program is more work for some people, but I don't think a city is going reduce its waste unless it's forced to do so.
> (J. Z., Hosford-Abernethy)

The quote alludes to the attitude–behavior gap; people don't change behaviors just because they are aware of a problem or even a solution. Behavioral change often requires system-level change to discourse and material structures. Making composting easy and reducing the amount of space for trash to landfills removed a barrier to positive behavior and reinforced a barrier to wasteful behavior.

Adapting to structural changes is challenging for some consumer households. Another *Oregonian* survey respondent embodies the struggle of adjusting to less frequent trash service: "We are a household of five with three school age children and we have a fair amount of trash, even though we are dedicated recyclers, and our garbage can is absolutely stuffed at the end of two weeks" (M. K., North Portland). Another writes:

> I'm way into reducing our footprint on the planet – we built a LEED-certified office building, I'm on my third Prius and have had a dinky garbage can for years. BUT, this new system has no grace! If our busy family misses the garbage week, then we have FOUR weeks between pickups. Not reasonable for a household with teenagers, pets and working parents. Yesterday, we gave up and ordered the biggest garbage can. It won't be full, when it gets out on time, but at least we won't have an unsightly overflowing can when the boys forget to roll it out on the "right" week.
>
> *(H. W., Dunthorpe)*

These citizens make efforts and even go to extra expense to comply with the program even though it has created some hardship for them. Others wilfully subvert it in shows of civil disobedience. Writes one such *Oregonian* survey respondent:

> We do not like the program. Not surprising, since it was the Emperor's (the Mayor's) idea. One size does not fit all. Why would we like a program, which costs the same but includes half the service? We have always composted. We have little in the way of food waste. That is not the issue. One size does not fit all. The extra week's garbage goes into the (Mayor Sam) Adams Bag. This is a single use, disposable plastic grocery bag (we have lots of them). It contains all waste, recyclable or not. It goes to the public garbage cans on the street; the specific street depends on where I'm driving that day.
>
> *(C. B., Cathedral Park)*

Another public commenter echoes the subversive tone:

> The easiest way to deal with the new trash pickup schedule is to dump your excess trash at gas stations, Fred Meyer, or those convenient trashcans that dot downtown. Maybe we can sponsor a "bring your trash to work day" every week to deal with excess trash?
>
> *(Mannyvel)*

Some citizens report that the new constraints on trash collection cause them to be more aware of their product acquisition behaviors, in particular as they relate to non-recyclable packaging. One citizen writes: "Our family makes active choices to reduce our garbage production, and the new city program helps us do that. It would be tragic to go back to the old way" (A. Z., Brentwood-Darlington). The schedule change of trash pickup from weekly to every other week facilitates awareness of garbage production and enables positive change towards reduction of landfill waste. Reduced waste begins at the point of purchase, and awareness of reduced trash capacity follows people into the marketplace.

Emotional impacts among the citizenry run the gamut from enthusiasm and civic pride to anger and protest. As an example of the former, one respondent writes, "I love the composting program! It makes me proud to be a Portlandia gal!" (T. G., Creston-Kenilworth). Another writes: "My little urban yard is too small to absorb the amount of compost I would make if I composted it all myself, and I really like the idea that these scraps are going to a useful purpose rather than a landfill" (C. B., Laurelhurst).

Prior attitudes toward City Government seem to be reinforced at both ends of the spectrum. Those that like the relatively progressive government view the program as an indicator of vision and strength. Those with more libertarian leanings see it as a sign of increasing government intrusion on their lives. Comments such as the following are illustrative:

> As typical with so many initiatives in Portland a significant new program is rolled out without adequately examining the consequences.... The food waste program is just the latest example.
>
> (J. P., Roseway)

> Every time I empty my food-waste container and have to clean up the mess, I think about Mayor Adams and how the city tried to recall him twice. This program looks like Mayor Adam's revenge.
>
> (W. A., Hillsdale)

Similar shifts appear to be happening in awareness and attitudes toward waste production and composting. Some Portlanders have expressed desires to receive free compost for their gardens, and this has been granted periodically. Others report real logistical challenges related to every other week trash pickup and resort to "cheating" (e.g. placing non-recyclables in their large recycling bins) and resulting guilt. Levels of overall environmental concern seem to be mostly unaffected. The curbside composting program has affected enhanced awareness of waste production streams.

Destabilizing and restabilizing actors

Certain non-human actors have emerged as especially vexing in their effects on the program. Consider the concerns of one commenter from the Belmont neighborhood:

> Even for those who can and do compost everything that is compostable, there is a substantial portion of household waste that is a disease risk but not

> compostable. Among these items are infant and adult disposable diapers, feminine sanitary napkins and tampons, absorbent, plastic-coated meat packaging and pads used in the meat departments of most commercial groceries, used kitty litter and animal waste from pet owners, etc. I suspect that some of these items are present in the household waste of most households in the City of Portland. In the warmer summer months, having these items sitting for two weeks will both make the city smell and allow for the spread of disease-carrying waste. I do not want to live in a city where walking down the street on a summer afternoon makes me gag.
>
> *(Belmont Folks)*

One of the most persistent problems arising from the program and a good example of the agency of non-human actors is the role of disposable diapers. This quote illustrates a common problem:

> We have always composted at our house and we like the idea of weekly compost and yard debris pickup. However, we really wish we still had trash pickup every week, too. We have a baby in diapers and another one on the way. Two weeks is just too long to go in between pickups.
>
> *(A. S., Kerns)*

Primarily this is a problem for families with infants or small children, but it also affects caretakers of the elderly in their own homes. Disposable diapers can't be composted or recycled, so they must go in the trash. Their bulk, however, tends to quickly fill trash barrels, which are now being picked up only biweekly. If the trash is full and the pickup is still days away, the continued flow of diapers forces the homeowner's hand. Some convert to cloth diapers. Some buy bigger trash bins and pay more for trash collection (Edwards 2011). Others scavenge space in neighbors' bins; yet others make clandestine diaper drops in commercial dumpsters or trash bins in places such as public parks (Slovic 2012a). One of the most pernicious problems from the city's perspective occurs when homeowners secret the diapers into any of the recycling bins. Diapers jam sorting machinery and must be hand-picked from the stream of compostable materials or recyclables.

As we saw in the Belmont Folks quote above, other non-human actors that have the power to destabilize the assemblage include the reality or the fear of pests, odors or disease that may insinuate themselves at the household level. These issues also tend to be solved at the household level, but often with the help of products from the marketplace. For example, compostable plastic bags help to contain and isolate odors and mess providing a solution for those who composted without a liner for their bin. The market for these bags emerged practically overnight in Portland.

Odors are also powerful actors at the neighborhood level. One consequence of the success of the program has been the delivery of more compostable material than processing plants could handle in a timely manner. Overburdened facilities began to smell, and that set neighborhood associations and activists into action.

In the short term, this necessitated shipping waste to other plants. Excerpts from a story in *The Oregonian* explain:

> Food waste from Portland's 6 month old composting program is traveling hundreds of miles to Washington because of competition between two composting companies .... Why so far? Metro works with two companies ... to manage two large hubs for transferring food waste mixed with yard waste .... The only nearby facility licensed to collect the mixture is [Company A] in North Plains; state regulators capped the amount of compostable material it could accept after nearby residents began complaining about the facility's smell .... All of the compostables from Metro South are sent to a compost facility north of Corvallis. But that facility is owned by [Company B] ... [Company A] won't send waste to its competitor's compost lot.
>
> (Slovic 2012c)

In the longer term, composting companies have built their capacity. As reported by Oregon Public Broadcasting (Sept. 13, 2012): "[Company A] has invested a lot of money improving its technology to control odour problems laying asphalt to unimproved surfaces, installing an intricate blower system."

As one might conclude from the public discourse from which we derive much of our data, the press has also been an influential actor in the course of the program, alternately destabilizing and restabilizing the overall assemblage of the program and its stakeholders. A barrier to obtaining our first interview with City of Portland officials was that they were still stinging from their relations with the press since program rollout. The fear that we might also feed information to the press made them initially wary of agreeing to an interview. After careful explanations of the purpose and scope of our research project, DPS leaders agreed to an interview.

In this chapter we are unable to plumb all the intricacies of the many relations and resulting translations to the Portland composting program. We can report that the program remains in effect at the time of writing and appears to be under no threat of failure. The relative stability of the program results in part from widespread public support and in part from additional mobilizations of technological and discursive actors from DPS, trash haulers, compost processors and the press, to name a few of the most important.

ANT lessons for consumer culture research

There are a number of reasons for consumer culture scholars to pay more attention to ANT and material agency. Perhaps the most important is that it helps us to see into certain blind spots in our field of vision. In embracing social constructionism and prioritizing consumer agency, we have systematically ignored the agency of actors such as nature, technology and infrastructure in the creation, reproduction, maintenance, destabilization and evolution of consumer culture. For all our interdisciplinary endeavor, one of our blind spots in consumer culture research is the potentially important and almost entirely neglected role of the natural and physical sciences in understanding consumption. In rejecting ontological and epistemological approaches borrowed from the physical sciences as inappropriate to studies of the social, we may have closed our eyes altogether to the value of those sciences. Of consumer researchers, consumer culture theorists are arguably the most inclined to consider holistic system perspectives when investigating consumption phenomena. And yet we would argue that to the extent that we ignore the relations among people, markets and the biosphere we fall dangerously short in our attempts at holism and system-level thinking. ANT underscores the realities that (1) all consumption is translation, (2) all consumption is connected and, (3) all of it connects back to the biosphere. We need to bring natural and physical sciences into the interdisciplinary fold to fully grasp the implications of these connections.

The call for uniting social and physical sciences already echoes in studies of human evolution and ecology (Mysterud and Perm 2007). Ehrlich and Holdren (1971) drew the connection between consumption and technology in their well-known formula $I = P \times A \times T$, wherein environmental impact (I) is understood to be the product of total population size (P), per capita affluence (A) or resource consumption, and the impact of the technologies (T) used to supply that which is consumed. Central to the equation is consumption, and not the mere fact or quantity of it, but also the modes of production that support and drive it. Those of us who study consumption and culture as our central professional pursuit have something to offer a world that is overheating and a society that is actively undermining its own life support system.

One important lesson from the Portland project is that for large-scale behavioral change for the good of society we should not be looking merely, or even primarily, for idealist solutions such as elevating consumer awareness and changing attitudes toward greener practices. Yes, attitudes shifted among many Portlanders, but the attitudes followed the behaviors more than the other way around. Waste has always been collected at the curb. The secret is enabling the behavior of moving household organic waste (e.g. food scraps, pizza boxes and soiled paper product) from the trash bin to the yard trimming and compost bin. And the real payoff is in behaviors that create massive reductions in organic waste to landfills and its conversion to a usable product. Changing the behaviors was a matter of making strategic changes to infrastructure and technology and supporting those changes with appropriate discourse.

References

Akrich, Madeleine, Michel Callon and Bruno Latour (2002), "The key to success in innovation part I: The art of interessement," *International Journal of Innovation Management*, 6 (June), 187–206.

Appadurai, Arjun (1986), "Introduction: Commodities and the politics of value," in Arjun Appadurai (ed.), *The Social Life of Things: Commodities in Cultural Perspective*, Cambridge: Cambridge University Press, 3–63.

Bicycling (2012), http://www.bicycling.com/ride-maps/featured-rides/1-portland-or (accessed February 2, 2012).

Çalışkan, Koray and Michel Callon (2010), "Economization, Part 2: A research programme for the study of markets," *Economy and Society*, 39 (February), 1–32.

Callon, Michel (1986), "Some elements of a sociology of translation: domestication of the scallops and the fishermen of St. Brieuc Bay," in John Law (ed.), *Power, Action and Belief: A New Sociology of Knowledge*, London: Routledge, 196–223.

City of Portland, Bureau of Planning and Sustainability, http://www.portlandonline.com/bps/index.cfm (accessed July 12, 2009 and May 22, 2014).

DeLanda, Manuel (2006), *A New Philosophy of Society: Assemblage Theory and Social Complexity*, London: Continuum.

Deleuze, Gilles and Felix Guattari (1988), *A Thousand Plateaus: Capitalism and Schizophrenia*, London: Athlone.

Dolphijn, R. and I. van der Tuin (2012), *New Materialism: Interviews & Cartographies*, Ann Arbor, MI: Open Humanities Press.

Edwards, Art (2011), "Portlanders get food buckets for composting," *kgw.com*, October 17. Available online at http://www.kgw.com/news/Portland-residents-get-food-scrap-buckets-for-new-compost-program-132023353.html (accessed September 4, 2014).

Ehrlich, Paul R. and John P. Holdren (1971), "Impact of population growth," *Science*, 171 (March), 1212–17.

Gell, Alfred (1992), "The technology of enchantment and the enchantment of technology," in Jeremy Coote and Anthony Shelton (eds), *Anthropology, Art and Aesthetics*, Oxford: Clarendon Press, 40–63.

Har, Janie (2010). "Portland Mayor Sam Adams names test neighborhoods for food scrap composting project," *OregonLive.com*, April 13, blog.oregonlive.com/portland_impact/print.html?entry=/2010/04/portland_mayor_sam_adams_names (accessed September 4, 2014).

Harris, Marvin (1979), *Cultural Materialism: The Struggle for a Science of Culture*, Walnut Creek, CA: AltaMira Press.

Kilbourne, William and Gregory Pickett (2008), "How materialism affects environmental beliefs, concern, and environmentally responsible behavior," *Journal of Business Research*, 61 (September), 885–93.

Kollmuss, Anja and Julian Agyeman (2002), "Mind the gap: Why do people act environmentally and what are the barriers to pro-environmental behavior?" *Environmental Education Research*, 8(3), 239–60.

Latour, Bruno (1987), *Science in Action: How to Follow Scientists and Engineers through Society*, Cambridge, MA: Harvard University Press.

Latour, Bruno (1999), "On recalling ANT," in John Law and John Hassard (eds), *Actor Network Theory and After*, Oxford: Blackwell, 15–25.

Latour, Bruno (2005), *Reassembling the Social: An Introduction to Actor-Network Theory*, Oxford: Oxford University Press.

Law, John (1988), "The anatomy of a sociotechnical struggle: The design of the TSR2," in Brian Elliott (ed.), *Technology and Social Process*, Edinburgh: Edinburgh University Press: 44–69.

Law, John (2004), *After Method: Mess in Social Science Research*, London: Routledge.

Law, John (2008), "Actor network theory and material semiotic," in Bryan S. Turner (ed.), *The New Blackwell Companion to Social Theory*, Hoboken, NJ: Wiley-Blackwell.

Martin, Diane M. and John W. Schouten (2012), *Sustainable Marketing*, Upper Saddle River, NJ: Pearson Prentice Hall.

Martin, Diane M, and John W Schouten (2014), "Consumption-driven market emergence," *The Journal of Consumer Research*, 40 (February), 855–70.

Miller, Daniel (1998) *Material Cultures: Why Some Things Matter*, Chicago: University of Chicago Press.

Mysterud, Iver and Dustin J. Perm (2007), "Conclusion: Integrating the biological and social sciences to address environmental problems," in Iver Mysterud and Dustin J. Perm (eds), *Evolutionary Perspectives on Environmental Problems*, New Brunswick, NJ: Transaction Publishers, 281–96.

The Natural Step http://www.thenaturalstep.org.

Oregon Public Broadcasting (2012), September 13, http://www.opb.org/news/article/still-holding-your-nose-portland-survives-one-year-of-curbside-composting/ (accessed September 4, 2014).

Palmås, Karl (2007), "Deleuze and DeLanda: A new ontology, a new political economy?" Paper presented 29 January at the Economic Sociology Seminar Series, the Department of Sociology, London School of Economics & Political Science.

Slovic, Beth (2012a), "Fines could be on the horizon for Portland's curbside food composting," *The Oregonian*, June 7. Available online at http://blog.oregonlive.com/portland_impact/print.html?entry=/2012/06/fines_could_be_on_the_horizon.html (accessed September 4, 2014).

Slovic, Beth (2012b), "More than 500 readers respond to survey on Portland's curbside composting," *The Oregonian*, February 15. Available online at http://www.oregonlive.com/portland/index.ssf/2012/02/over_500_readers_respond_to_su.html (accessed September 4, 2014).

Slovic, Beth (2012c), "Some Portland composting travels to Washington: Portland City Hall roundup," *The Oregonian*, May 10, http://blog.oregonlive.com/portlandcityhall/2012/05/some_portland_composting_trave.html (accessed September 4, 2014).

Thomas, Tandy Chalmers, Linda L. Price and Hope Jensen Schau (2013), "When differences unite: resource dependence in heterogeneous consumption communities," *Journal of Consumer Research*, 39 (February), 1010–33.

Thompson, Clare (2009), "The 15 most sustainable U.S. cities," *Grist*, http://grist.org/article/2009-07-16-sustainable-green-us-cities/full/, (Accessed February 2, 2014).

Van Binsbergen, Wim (2005) "Commodification: Things, agency, and identities: Introduction," in Wim MJ Van Binsbergen and Peter Geschiere (eds), *Commodification: Things, Agency and Identities: The Social Life of Things Revisited*, Münster: Lit Verlag, 9–51.

Witt, Matthew (2004), "Dialectics of control: The origins and evolution of conflict in Portland's neighborhood association program," in Connie P. Ozawa (ed.), *The Portland Edge: Challenges and Successes in Growing Communities*, Washington, DC: Island Press, 84–101.

7

CLOTH LOOP

An attempt to construct an actor-network

Eva Gustafsson, Daniel Hjelmgren and Barbara Czarniawska

Introduction

Cloth Loop is a project launching a new approach for collecting used clothing and textiles and the focus of our study. Cloth Loop was started by the management of Discount Store Inc. (DSI),[1] a large, popular department store in V City in Southern Sweden, in cooperation with one of Sweden's larger humanitarian organizations, Global Aid.

In concrete terms, Cloth Loop is a collection facility where nearly new and well-worn clothes are deposited for recycling. It is located about 2 km from DSI, and consists of a 300 m^2 warehouse, staffed every weekday between 08.00 and 18.00. Clothing and other textiles received from donors were to be sorted and distributed by Global Aid that would use them for financing its various aid projects. Actually, this is how Cloth Loop was supposed to work, but the route to that end proved somewhat complicated. We present that process using Michel Callon's (1986) and Bruno Latour's (1987) version of actor-network theory to show how various *actants*[2] and actors were recruited to form a network around Cloth Loop. We begin by sketching the context of this project, describing current systems for dealing with used clothing. We continue with a chronological narrative of the way Cloth Loop came into being, and end the chapter with an analysis of the process, attempting, in the conclusion, to explain its shortcomings and suggest some solutions.

Textile waste – an increasing problem in Western societies

Because Sweden's consumption of clothing is high, the amount of used clothing is steadily increasing. In a recent study, Swedish Environmental Emissions Data reported that the net inflow of clothing and textiles in Sweden during 2008 was 131,800 metric tons, which corresponds to almost 15 kg per person (Carlsson *et al.* 2011). The net inflow of garments and home textiles increased by nearly 40 percent between 2000 and 2009. Increased consumption of textiles is evident in many other countries, and there are several reasons for it: higher welfare benefits, lower clothing prices, and more rapid fashion swings. Many of the major clothing stores launch four collections per year, rendering the garments "old" just a few months after they were purchased.

Approximately 7,400 metric tons (8 kg per person) of used textiles are thrown in household trash bins and converted into energy in district heating plants (Tojo *et al.* 2012). District heating is based on economies of scale; a large system delivers heat to many users in a specific geographical area. Compared to smaller-scale options, district heating plants are more efficient and consume less fuel, yielding economic and environmental benefits. As of this writing, there are 30 waste incineration plants in Sweden, and the number continues to rise (Vowles 2013). From an environmental standpoint, only landfill is a worse alternative, because it results in the leakage of toxic substances, and organic material produces methane gas (ibid.). According to the European Union's "waste hierarchy",[3] a more environmentally friendly and economical alternative to burning textiles is reuse (using according to their original function) or recycling (destroying the product design and using the material for the manufacture of new products, such as sound insulation in automobiles, stuffing in mattresses and furniture, clothing, and fabric for new clothing).

Ways of handling used clothing

Most of the clothing collected for reuse or recycling in Sweden is currently collected by various non-governmental organizations (NGOs) – Human Bridge, *Myrorna*, the Red Cross, *Emmaus*, *Stadsmissionen*, and *Humana*, for example. In 2008, the eight largest NGOs received approximately 26,000 metric tons of clothing and textiles – 3 kg per person per year (Carlsson 2011). What cannot be submitted directly to thrift stores is collected in containers found in various shopping centers and community recycling sites. Some organizations, such as *Stadsmissionen*, pick up clothing at the donor's home. Donated textiles are usually shipped to a central collection point, where usable items are sorted, and reconditioned if necessary, prior to their distribution to various charitable organizations and thrift stores.

Of all the clothes collected by the NGOs, only 3,000 metric tons (0.3 kg per person) are sold in Sweden. The majority goes to various purchasers, particularly in Eastern Europe, or to the big recycling companies in Germany, Holland, and the Baltic States (Tojo *et al.* 2012). About 19,000 metric tons (2.1 kg of used textiles per person) are exported. The clothing sold to Eastern Europe is used in that local market or re-exported – to Africa, for example (Tojo *et al.* 2012). Human Bridge renovates and carries medical equipment and medicines to Third World countries and donates textiles directly to the needy in disaster areas. Other NGOs, such as *Stadsmissionen*, donate clothes to homeless and other vulnerable people in Sweden.

Over the past decade, interest in textile recycling has increased, resulting in some new ventures. Several Swedish companies, including *Houdini* and *Fjällräven*, are part of Eco Circle, the recycling project initiated by the Japanese fabric manufacturer, *Teijin*. *Teijin* began by recycling polyethylene terephthalate bottles, but has subsequently expanded its operation to include the recycling of fleece, which was initially restricted to Japanese school uniforms and gym clothes. The project's big break came in 2005, when Patagonia, a major US manufacturer of outerwear, decided to join the project. In 2009, 95 percent of the Swedish stores that sell Patagonia products were equipped with recycling barrels where customers can leave their old fleeces (Hagberg 2009).

Systems for recycling have recently been adopted by fast-fashion companies. In 2012, H&M launched a recycling program in Switzerland, whereby customers are encouraged to donate their old clothes to the H&M stores chosen for the test. By late 2013, H&M's goal of having all its stores accept used garments from their customers (Cosnier 2013) was realized. For each bag of used clothing donated, customers now receive a voucher that they can use for buying new clothes. H&M collaborates with I:Collect, which is responsible for finding the best possible use for the collected material.

There has been growing interest in textile redesign in recent years, resulting in such new ventures as the *Zäntrumprojektet*, headed by the Swedish Red Cross. *Zäntrumprojektet* started in 1983 with the goal of teaching unemployed women to sew and eventually become dressmakers. The project currently addresses young people with difficulties entering the labor market; they manufacture quilts, home textiles, and decorative items that are sold at Red Cross bazaars (Swedish Red Cross 2012). Even the fashion industry has shown some interest in redesign. In 1994, fashion designer Jean Paul Gaultier presented a small collection of coats, jackets, skirts, and pants (trousers) made from old jeans (Lotti 1994). Swedish clothing companies, too, have shown an interest in redesign. In 2009, Lager 157 introduced a collection of "new clothes" – used garments redesigned beyond recognition by Amanda Ericsson, who also runs her own fashion brand: Dreamandawake (Lager 157 2009). Most actors involved with redesign are small businesses like Dreamandawake, with usually one or two designers.

Anti-programs – competitors of Cloth Loop

Actor-network scholars speak of *anti-programs* when there is more than one attempt to construct networks dedicated to similar activities. Various systems for textile recycling exist, aimed at reducing the proportion of clothes thrown in household trash. There are already several smaller and larger actor-networks in the field, each with its own action programs, which can be seen as anti-programs to that of Cloth Loop. The survival of all these actor-networks depends partly on their ability to make these systems both environmentally and economically sustainable. To date, there are no recycling facilities in Sweden. Its recycling is sent to other countries, which means that transportation costs limit the environmental and economic benefits of recycling. In addition, there are few industries in Sweden that use recycled materials. The market for reused clothes in Sweden – whether redesigned or vintage – is small (Ekström *et al.* 2012).

The construction of an actor-network begins

The construction of the actor-network, Cloth Loop, began with a research application that a professor at the University of Borås wrote to The Trade Development Council, in Spring 2010. The Council had announced funds for research on sustainable trade with the goal of finding solutions to increase social and ecological sustainability. The application was first sent to the CEO of DSI asking if they wanted to cooperate. In addition to making it possible to study the growth of a specific solution in real time, DSI was the first actant to be recruited into the actor-network that was to be built.

How can DSI contribute to sustainability?

It was not clear initially what the textile recycling project should comprise, other than that it should address the collecting of old clothes and solutions for increasing sustainability – generally, and within DSI in particular. Planning for the project began at a meeting at which two of the authors met six DSI representatives to address the question of what to do with the clothes that were to be collected. As a group of researchers at the University of Borås studied how old clothes could be recovered as biogas, it was decided to explore this option. DSI already had a plant for the production of biogas from organic matter, so it seemed obvious that this should be the first possibility to explore.

At the second meeting in January 2011, the research team that now included two US scholars, Diane Martin and John Schouten, met with the CEO of the department store and the three area managers whose work would be affected by the proposed project. The US colleagues held a presentation on Wal-Mart's efficiency improvements that focused on sustainable development without increased cost. They concluded that Wal-Mart's sustainability work had benefited from the strong conviction of its management team, which took sustainability issues seriously enough to risk the ruination of a highly successful business model.

At the third meeting in February 2011, DSI was represented by the store's corporate social responsibility (CSR) manager and the person responsible for environmental issues. DSI representatives presented the store's overall efforts at reducing energy consumption and resolving environmental and social issues at the manufacturers. The DSI representatives maintained that the environmental and social aspects of the project must not be "mediocre" – that all partners must be legitimate and well known. We were reminded that everything DSI does is on large scale, so if collected clothes are to be converted into biogas, it must be at a large facility. Unfortunately, the biogas research project at the University of Borås was still in the laboratory stage, so there was no large-scale solution that could be implemented at that time.

Another objection to the biogas solution, this one from DSI representatives, was that customers might dislike the idea of turning their clothes into biogas, preferring to put them to use in other ways – donating them to people in need, for example. Moreover, it would take at least three years before a commercial solution would see the light and, in the meantime, the pilot plant had to be placed in Borås, with the research and necessary equipment. This meant that the clothing would have to be transported to Borås after sorting. Furthermore, no matter what portion of the textiles was converted into biogas, DSI would require a sorting facility, as they did not have the facility or the skills to sort them themselves.

DSI finally decided not to explore the biogas option further. From then on, the project focused on reuse of the used clothing, and the NGOs from the already existing recycling network (Ekström and Salomonson 2014) – the charitable aid organizations – were brought in for consultations. Another actant had to be recruited.

Aid organizations are invited to join

It turned out that all three charities from the network created in an earlier project (Ekström and Salomonson 2014) – the small NGO Local Aid and two large NGOs, Global Aid and Swedaid – were interested in participating. The first meeting took place in May 2011, beginning with a discussion on the scale of the project. DSI representatives doubted if the NGOs had sufficient resources to accommodate all the clothing that would be donated. As it turned out, this concern was unwarranted. The two larger NGOs, Global Aid and Swedaid, had sorting facilities in Sweden that could accommodate large quantities of clothing. Global Aid was able to send clothes to large-scale processing facilities in Central Europe, and the demand for used clothing on the world market was greater than the supply. A large-scale project was only a problem for Local Aid, the smallest of the three NGOs, as their system was based on local solutions.

The worst-case scenario was that too few clothes would be donated, and various ways were considered to encourage customers to bring their old clothes when shopping at DSI. DSI wanted to avoid any form of monetary reward, mainly because it would result in additional costs. For them, the goal was to achieve a break-even; the project should generate neither revenue nor additional cost. Also, a previous study (Ekström *et al.* 2012) had revealed that only 9 percent of DSI customers reacted positively to the idea of obtaining monetary rewards for donating their used clothing. Various other types of rewards, such as pins that said "I have donated clothes to DSI" or tokens that could be used instead of coins for the store's shopping carts, were therefore discussed.

The discussion then moved to what should be collected. Swedaid was interested in collecting other used items, such as books. Was it possible to locate a second-hand shop on the DSI site, run by the NGOs with or without the help of DSI? It turned out that DSI's management team had already discussed that solution, and that it would likely be received positively by its board. At this point, Swedaid's representatives pointed out that a second-hand shop would require a larger budget than the one needed to establish a collection facility for used clothing. In addition to the facilities and the staff, a second-hand store required on-site staff to sort the goods and strategies for dealing with unsold items.

After the first meeting with the three aid organizations, the project group had a rough idea of what might work. DSI representatives would have preferred a more innovative solution than merely collecting clothes, but the options had proven too costly or too logistically complicated. At the next meeting, the group decided, the focus should be on the design of a collection facility at DSI.

The second project meeting was held in June at DSI, and the representatives of the three aid organizations were shown two possible sites for the facility. One was a large gravel field, formally used for football, with space for containers, close to the road that many DSI customers take. The other possible location was within a fenced area around the DSI warehouse, but some kilometers away from the store. Still, the fence meant that the containers would be protected. The group also decided that the recycling station must be equipped with facilities for manual sorting.

After the tour of the potential sites, DSI representatives presented the requirements the aid organizations had to fulfill. The first was the possibility for the donors to earmark clothes for specific purposes. Furthermore, the aid organizations would have to control the value chain, so the project would not risk a negative effect on DSI's goodwill. It was also desirable that customers be kept informed about the amount of clothes collected.

This project had several drawbacks for the Local Aid; as the smallest NGO, it already had too many clothes to sort and no interest in receiving more. Yet participation in the project represented an opportunity to reach out to more people who might be willing to donate money, and there was a hope of eventually opening a thrift store in V City. As the project was currently framed, however, it was not attractive to Local Aid. Although the DSI representatives pointed out several benefits, such as expanding Local Aid's activity outside the big cities, the disadvantages were more numerous than the advantages.

Local Aid was a valuable partner for DSI, because it had more contacts with local charities than the two larger NGOs had. One solution was for Local Aid's members to drive to DSI each month to choose and collect some part of the donated clothes. This plan meant, however, that the most saleable clothing would be chosen by the smallest NGO, leaving the less desirable pieces for the two larger charities. It turned out to be difficult to include both large and small organizations in the project.

The two larger NGOs appreciated the monetary value of the clothes collected. With the money from the sales, Global Aid planned to pay for the transfer of old hospital equipment to developing countries. The other large NGO, Swedaid, had recently built a new sorting facility, and the pilot project could help to utilize that resource better.

The meeting ended with an invitation to visit the aid organizations' facilities. The visits took two days, during which time the DSI people learned a great deal about the work of aid organizations. They had previously shared a common misconception that all the used clothing collected went straight to disaster relief or to the needy in Sweden. Yet Local Aid does not even export clothing; it converts the majority of collected goods into money, which it uses for other charitable purposes. The two larger NGOs sell a large part of the collected clothing abroad. The DSI representatives realized that this information had to be communicated to their customers.

Time out

In August 2011, the group met again to decide the timetable for the pilot project and the division of labor. DSI announced two decisions: to locate the facility on the former football field (first option discussed earlier) and to limit their fixed investment to a fence around the site. A drive-through tent for depositing clothes and barracks for staff would be hired. DSI's budget also included one full-time employee, but because that person could not be on site 24 hours a day, every day, they asked the aid organizations for support with on-site staffing. Global Aid offered to help.

It turned out that the two larger NGOs did not use the same type of containers, which would seriously hamper cooperation. At the end of the day, only Global Aid could promise to participate without first consulting their colleagues.

Then the project stagnated for a while. The first to pull out of the project was Local Aid, which simply did not have the resources required. Swedaid dropped out a month later, for less obvious reasons. The only organization that remained positive about the project was the other large NGO, Global Aid. But now it was the DSI people who hesitated. They had heard about similar pilot projects, and were not interested in "being No. 2."

Restarting

The project was given a new life when a new purchasing manager started her job at DSI. The project had been introduced to her as one of many during the handover of her duties. After reading about it in greater detail, she contacted people engaged in a similar project, and concluded that the projects were not competing. She established a working group with the CSR manager as project manager with responsibility for communication and marketing; another person, in charge of environmental issues, was to solve practical problems like contacting the appropriate contractors and completing the necessary permission forms.

Global Aid was then re-contacted, and in November 2011 a new meeting took place. The most conservative expectation was that 2 percent of DSI's customers would each bring 3 kg of clothing to the site. For practical reasons, the start date for a pilot project was set for the middle of May 2012. The plant was to be established with a minimum of fixed investment, but it would be possible to scale it up rapidly if it attracted more customers than planned. Global Aid would provide the staff, DSI would help with the preparation of the plant, and the agreement between the two organizations was signed in December 2011. DSI was responsible for all investments and the NGO was responsible for operations, the personnel, and, when necessary, new trucks and containers.

When all the costs of the planned pilot were known, DSI's investment equaled over 500,000 Swedish Krona – too heavy an investment for a project that might end in six months. It was therefore decided in February 2012 to use the existing facility (the second option discussed before), which already contained most of what was needed: fences, security, electricity, heating, toilets, and a kitchen. The investment required to start the facility running was one-fifth of the cost of the alternative. The facility provided nearly 300 m^2 of storage, which could be extended to 800 m^2. This meant that the business could be run entirely indoors.

Eventually, Global Aid had three containers in place. The staff packed the incoming garments in plastic bags, weighted them, and placed them in containers. Cloth Loop was to be open from 08.00 to 18.00 on weekdays and from 09.00 to 15.00 on weekends, and at other times clothes could be left in the collection containers placed outside the gate. Staff would empty these containers in the morning and go through the same procedure of packing, weighing, and repacking. The collection containers that Global Aid had placed in other locations in the region would be emptied and repacked at Cloth Loop that would now become the collection hub for the whole region.

The marketing of Cloth Loop

The marketing of Cloth Loop began a few weeks before its opening on 15 May 2012. It was important not to "beat the big drum," and present Cloth Loop as bigger or more important than what it was – not as "the savior of the environment," but as a service that DSI's customers and others could use to dispose of old clothes in an environmentally sound and ethical way. DSI did not want to play the role of educator to their customers, and the marketing was not to sound dry and bureaucratic. All information about Cloth Loop, it was decided, should be as straightforward and simple as possible: Customers would learn that there was now an opportunity to donate their old clothes when they were shopping at DSI. And as the purpose of Cloth Loop was to reduce the amount of clothing and textiles that end up in municipal waste, it was important to communicate clearly that DSI also accepts well-used clothes.

DSI's vision, written by its CEO, was quoted in an article written for DSI's monthly magazine (which ran to approximately 200,000 copies) by an advertising agency, based on facts provided by DSI's marketing department. It included hours of operation, directions, the purpose of Cloth Loop, and a kindly worded invitation to donate clothes. Cloth Loop was also presented on the DSI website, the DSI blog, Facebook, Twitter, and on various signs around V City. The website informed the reader about the process of the collection, and those who wanted to learn more were referred to the Global Aid website. There was a facade sign at DSI, a road sign showing the way to the site, and posters on the collection containers that explained which clothes were accepted.

Cloth Loop was marketed on several billboards in the store and on the TV monitors mounted in the areas where the customers tended to slow down – near the toilets, the restaurant, and the cafés. In addition, customers could pick up a flyer about Cloth Loop at the exit. In all the advertisements it was stated that those who deposited clothing would receive a voucher for coffee that could be redeemed in any of the store's coffee places.

Previous experience suggested that the chosen marketing channels were effective. Thus to avoid "traffic chaos and possibly bags with old clothes thrown all over the place," as one of the DSI representatives put it, it was decided to ration the amount of advertising for Cloth Loop. The idea was to increase the marketing slowly, to ensure that the plant survived an influx of customers donating their used clothing.

How to convince people to donate

Cloth Loop held its grand opening on 15 May 2012, but the only people present were the members of the research group and representatives of DSI and Global Aid. No customers queued to deposit their clothes and no media were there to cover the event. The opening was a disappointment to everyone involved. The press release had had no impact; a report on the local radio and a few items in local newspapers were the only results. This was unexpected, as the media had previously shown great interest in the research group's projects. Moreover, DSI almost always received a great deal of attention in the media when something new happened.

To understand the situation better, the research team decided to conduct interviews. We placed ourselves on the stairs leading to the restaurant, and interviewed 120 people. It turned out that the vast majority had not even noticed Textile Return. In late autumn, therefore, DSI decided to increase the marketing of Cloth Loop. Large signs were placed in the parking lot and additional collection bins were placed adjacent to the store's shopping-cart stall. It was reasonable to assume that DSI customers could not help but recognize Cloth Loop in the face of these efforts. Two weeks before Christmas, we conducted 47 further interviews with the DSI customers. Only twelve of them knew about Cloth Loop. Five had read about it in DSI's customer magazine, one had read about it on the website, and one had noticed the large signs in the parking lot. The remaining three did not remember how they had found out about it. Of the twelve who knew about it, four had brought used clothing and two had donated more than once. They all lived in rural areas a bit east of DSI, visited the store often, and had no other real shopping options. The 35 people who had not brought clothes to Cloth Loop told us that they regularly donated clothes to another charity organization. Proximity and convenience were the most common reasons for their choice of collection site; the reasons people gave for donating their used clothing to Cloth Loop, then, turned out to be the same reasons people chose *not* to leave clothes there.

Why is Cloth Loop not an actor-network yet?

The creators of Cloth Loop had high hopes for their creation, which failed to materialize – at least in full. To understand what happened, we analyzed the events with the help of Callon's (1986) four stages of translation, which he used to characterize the successful construction of an actor-network: problematization, interessement, enrollment, and the mobilization of allies.

Problematization: textile waste as an environmental hazard

The origins of Cloth Loop can be retold in several ways, and the narrative differs depending on where one is in the network. For us, Cloth Loop started as a time-limited research project on sustainable trade, funded by The Swedish Retail and Wholesale Development Council. The exact meaning of "sustainable trade" is debatable, but a common interpretation is extended manufacturers' responsibility, which means that the trading companies are responsible for all aspects of the value chain, from raw material to finished product.

The application submitted to the Council was based on research that pointed to the increasing amount of textile waste as a waste of resources, and thus an environmental problem – the opposite of "sustainable trade." As stated in the application, consumers and manufacturers must take joint responsibility for reducing textile waste. The purpose of the study was to generate knowledge on ways to accomplish this goal and the pilot project for recycling textiles in cooperation with DSI was one part of it.

Our application was the result of past relationships. A year earlier, the research team had conducted a study on DSI and established a contact with the department store's CEO. It is unclear who in this case recruited whom; the research team and DSI approached the problem of textile waste from somewhat different perspectives. The most accurate description might be that it was the problem description itself that connected DSI and our research team in this new project.

Interessement: who might be interested in our project?

According to Latour (1987), the way for an actant to obtain allies is to translate its goal into the goals of other actants to demonstrate their shared interests: "You cannot reach your goals straight away, but if you come my way, you would reach it faster, it would be a short cut" (Latour 1987: 111). Assets that can be used for reaching certain goals are the means possessed by the potential allies. One potential ally was the group of researchers in chemical engineering who conducted research on waste management with a focus on the conversion of textile waste into biogas. This research team was connected to the Borås municipality, which had invested in biogas as the primary solution for waste management. If the chemical engineering research group could connect to the pilot project, their goal, which was to transform textiles into biogas, would be significantly strengthened. Through DSI, the biogas network would be given access to the means such as facilities for processing waste into biogas and a suitable fundraising partner. The interessement failed, however, because the connection required investments that were well over the limit of the DSI management's budget.

It was decided that the pilot project should be attractive for those interested in the extended manufacturers' responsibility for the environment. One of these actors was a fashion retailer who had mobilized other retailers to create a functional recycling system for fashion stores. Because of the well-known brands that participated, this already existing network could significantly influence the design of the future system extending manufacturers' responsibility for the environment. Therefore DSI's hesitation: Should they terminate their project and join the others – a competing interessement? But the network of retailers wanted a small-scale recycling system with many collection points, one for each fashion shop. For DSI, a small-scale recycling system was not an option.

The senior researcher in our team was a member of a network through which DSI contacted the three aid organizations. Their goal, however, was not to reduce the amount of textile waste, but to obtain more money for their charity operations by increasing the amount of donated clothing. Because these three aid organizations had established systems for collecting clothing, they were able to translate the goal of the pilot project to be consistent with their own goals; collected clothes were recovered, and only rags were sent to incineration. It was not necessary that all the involved parties shared the same goals, as long as the necessary actions could be connected in such a way that the objectives of the pilot project were met. The interests of all parties did not have to be identical, but had to be aligned.

The key goal for DSI was to persuade its visitors to deposit their clothing when shopping – clothes that would otherwise have ended up in the household bin. To do so, DSI engaged their usual communication channels: the DSI website provided information about the collection site, and offered a link to Global Aid's website. The link to Global Aid demonstrated that Cloth Loop was the logistical solution necessary to deal with donated clothes in a good, charitable, and environmentally friendly manner. Through the DSI blog, customers could interact with other customers and convince them to bring their used clothing to Cloth Loop. DSI published a long article about the reasons customers should donate their clothes to Cloth Loop, emphasizing the green aspect of the donation. Finally, a large number of TV monitors and various types of signs were installed to increase visitor awareness of the collection site and the availability of clothing collection. In other words, they did practically everything they could. But enrollment and mobilization continued to be obstacles.

Enrollment: contracts and commitments

DSI's participation in the project was influenced by certain assumptions, one of which was that the project would finance itself. Cost awareness is a cornerstone of DSI's business model. A second assumption was that DSI should never try to teach its customers how they should live their lives. A third assumption was that DSI is not a leading player in the market for sustainable issues. Obviously, DSI brought these three assumptions into the project, while the aid organizations tried to insert theirs. Yet Local Aid and Swedaid did not manage to influence the pilot project in a direction that was consistent with their own goals, and dropped out of the project. Swedaid had an untapped sorting capability, but could not manage the large quantities of clothing that were expected to be donated in V City. In addition, they were interested in expanding the donated items beyond clothing to other articles such as books. Because the focus on textiles for the pilot project was not negotiable, the members of Swedaid lost interest.

The major goal of Local Aid was to reach more people, and DSI was intended as a showcase for their organization. One of the advantages that Local Aid brought to the negotiation was its tradition of providing clothing to the homeless. This local perspective was very much in line with the assumptions of DSI's representatives. On the other hand, Local Aid lacked the capacity for transporting and sorting the amount of clothes the donors were expected to deposit – small trucks for picking up donated goods, for example. And, unlike the other two NGOs, Local Aid owned no containers. So even if some of Local Aid's interests were aligned with the pilot project, there were other requirements that prevented it from becoming a part of the Cloth Loop actor-network.

The members of the pilot project succeeded in aligning their interests with those of Global Aid. Global Aid's goal in joining the project was to streamline its logistics and increase the flow of goods. It had an advanced logistics system that would considerably decrease DSI's operating costs. A guided tour of the charitable organizations' establishments showed that Global Aid had the resources needed to operate Cloth Loop. Furthermore, Global Aid had a strong interest in enrolling in the network. A key incentive was the opportunity to have a large-scale, staffed collection site in Southern Sweden, reducing transportation costs and costs for garbage handling. Cloth Loop also provided an opportunity to reach out to new donors – people living in rural areas, for example.

But the argument that Cloth Loop offered a convenient way to dispose of clothing appeared not to be sufficient for enrolling donors, who may have considered it easier to throw their old clothes in the household trash or give them to their usual charity. Moreover, even if the DSI customers learned about Cloth Loop, they could not do what they were asked to do until they had examined their wardrobes, chosen the clothes they wanted to donate, and returned to DSI. Consequently, Cloth Loop required more careful planning for the disposal of their used clothing. It is not surprising, then, that it was primarily DSI employees and people who lived nearby – and likely came to DSI repeatedly – who became enrolled, to use Callon's (1986) term. Additionally, while the other actors in the network – Borås researchers, DSI, and Global Aid – were already properly constituted actor-networks, the potential donors were dispersed actants who could not be collectively enrolled.

Mobilization: unresponsive consumers

Thus, the mobilization of "properly enrolled actors" around Cloth Loop was not a problem, but the mobilization of actants – customer donors – was. They had to remember the advertisements and act accordingly the next time they visited DSI. For many people, it would require a change of habit. According to a study by Ekström *et al.* (2012), older but expensive, stylish, and well-preserved clothes are likely to be sold second-hand, bartered, or given to friends and family members. Clothes of less financial value or with minor defects are mainly donated to charity, torn into rags, or thrown away. These habits render anti-programs preferable to Cloth Loop's program.

One such anti-program was offered by the local heating plant. The plant produces energy from waste that people throw in the garbage can. As mentioned, there are 30 waste incineration plants in Sweden, and their number continues to rise. The current situation, with heating systems that require a constant flow of garbage, might hinder the development of alternative systems for collecting used clothing and textiles. Another anti-program is being offered by the NGOs with collection containers located at various shopping centers and community recycling sites. Last but not least, even if DSI's customers did as they were asked, it would likely result in singular actions, possibly repeated sporadically, with a long time between donations. For most customers, disposal of clothes at Cloth Loop would never become part of their routine. It seems, then, that Cloth Loop is not likely to offer a serious challenge to its anti-programs.

Can anything be done?

If anything can be done to rectify this situation, it would still be within the interessement. DSI has a low public profile on environmental issues. The company is continuously working with various environmental issues, but environmental work has not yet achieved a place in the store's marketing. The green argument for submitting clothes to Cloth Loop was in the background of the advertisements, and the convenience argument was in the forefront. DSI customers who had already enrolled into other networks and deposited their clothes elsewhere had no need for a new logistical solution. The "green argument" should have been given priority for enrolling new donors to Cloth Loop. There is a deeply rooted tradition of the aid organizations to highlight the philanthropic motives for donating clothing, and this motive did not dominate the marketing for Cloth Loop. Within this tradition, people do not donate torn clothes, because torn clothes are not helpful to people in need. Hence, philanthropic motives will not make people stop throwing their ragged clothes in the garbage can. Global Aid has begun to highlight the green aspects of clothing collection, but their activity as a whole is still based on the notion that clothing is donated for charitable purposes – to give to the poor.

The media's lack of interest in this project was striking. To date they have shown very little interest in Cloth Loop, although DSI is still popular in the press, not least because of two reality shows based on the store and its employees. When DSI was described in one of the larger Swedish tabloids, Cloth Loop played no part in the presentation. Has it ended up in a media shadow? And if so, why? One possible explanation is that the public image of DSI does not match what is happening at the collection site. It seems that the media has created an image of the store where people go to "shop till you drop." Global Aid has begun to redefine their image by stressing sustainability, but the question is whether it will be as easy for DSI, which has often been portrayed as a symbol of consumption.

It is likely that other retailers who strive to integrate sustainability with consumption share this dilemma. Not all retailers have such a pronounced consumption profile as DSI, but global fashion brands like H&M certainly have the same problem. It may be seen as a paradox that fast-fashion retailers claim to work for sustainable development. Perhaps the retail business should focus on facilitating peoples' disposal of clothes and let other organizations convince people why it is important for them to change their disposal behavior. After all, it is impossible to ignore the fact that global consumption is increasing. There are new and rapidly growing consumer markets, particularly in Asia, and they will require the same supply of fashion products that the Western consumer has long taken for granted. From this perspective, every attempt to reduce the environmental impact is of crucial importance.

And finally, we want to add a practical solution for the mobilization of customers, as suggested by Diane Martin (personal communication, November 21, 2013). If DSI were to print on their shopping bags a text like "I will bring my old clothes in this one," possibly with a picture of Cloth Loop, it would serve as a strong memory aid the next time the customer came to DSI. As Callon and Latour would say, "Let the artifacts do the work!"

Notes

1 The names of organizations under study are fictive.
2 *Actants* in actor-network theory are units that act or are acted upon. Only if and when their actions succeed do they acquire "character" and become "actors" (Czarniawska and Hernes 2005).
3 http://ec.europa.eu/environment/waste/framework/ (last accessed 2014-01-27).

References

Callon, Michael (1986) "Some elements of a sociology of translation: Domestication of the scallops and the fishermen of St Brieuc Bay," in John Law (ed.) *Power, Action and Belief: A New Sociology of Knowledge*, London: Routledge & Kegan Paul, 196–233.

Carlsson, Annica, Hemström, Kristian, Edborg, Per, Stenmarck, Åsa, and Sörme, Louise (2011) *Kartläggning av mängder och flöden av textilavfall* (Mapping of the amount and flow of textile waste), Stockholm: SMED.

Cosnier, Maria. (2013) "Återvinning har blivit trendigt," *Miljöaktuellt*, February 22.

Czarniawska, Barbara and Hernes, Tor (2005) "Constructing macro actors according to ANT," in B. Czarniawska, and T. Hernes (eds), *Actor-Network Theory and Organizing*, Malmö: Liber, 7–13.

Ekström, Karin M. and Salomonson, Nicklas (2014) "Reuse and recycling of clothing and textiles – a network approach," *Journal of Macromarketing*, 34(3), 383-99.

Ekström, Karin M., Gustafsson, Eva, Hjelmgren, Daniel and Salomonson, Nicklas (2012) "Mot en hållbar konsumtion – En studie on konsumenters anskaffning och avyttring av kläder" (Towards a sustainable consumption – A study of consumers' acquisition and disposal of clothes), in *Vetenskap för profession (Science as Profession)*, Borås: Borås University College.

Hagberg, Gunnar (2009) "Fleece blir till fleece som blir till fleece som blir till fleece …," *Norrköpings Tidningar*, February 21.

Lager 157 (2009) "Kulturnatta gästas av turnerande bytesbutik," http://www.mynewsdesk. com/se/lager-157/pressreleases/ (accessed 3 September 2014).

Latour, B. (1987). *Science in Action: How to Follow Scientists and Engineers Through Society*. Milton Keynes: Open University Press.

Lotti, Ander (1994) "Jeans – ett minne blått?," *Expressen*, May 15.

Swedish Red Cross (2012) Annual Report.

Tojo, Naoko, Kogg, Beatrice, Kioroboe, Nikola, Kjaer, Birgitte, and Aalto, Kristiina (2012) *Prevention of Textile Waste – Material Flows of Textiles in Three Nordic Countries and Suggestions on Policy Instruments*, Nordic Council of Ministers.

Vowles, Kjell (2013) "Soporna som går upp i rök på en brännhet marknad," *Effekt och klimatmagasinet*, June 23.

PART III
Socio-cultural views on waste

8

EXPLORING FOOD WASTE THROUGH THE LENS OF SOCIAL PRACTICE THEORIES

Some reflections on eating as a compound practice[1]

Dale Southerton and Luke Yates

Introduction

Waste reduction is a substantive problem, and one that reflects broader societal challenges posed by the ideal of sustainable consumption. Consistent with dominant approaches to sustainable consumption, there is a tendency to treat waste either as a matter of production-side inefficiencies or the responsibility of consumers (either individuals or households) who need to be encouraged or empowered to adopt pro-environmental behaviours and lifestyle choices. Regardless of orientation, debate is firmly rooted in attempts to render today's 'normal ways of life' more efficient and less wasteful. As with broader debates about sustainable consumption, today's 'normalities' are rarely questioned or examined with respect to alternative arrangements for the organization of daily life. The consequence is a fractured, often domain or sector specific, and piecemeal approach to tackling sustainable consumption that fails to consider the inter-connections between the multiple processes and practices from which contemporary everyday lives are configured.

This chapter seeks to locate and explore food waste through the application of a practice theoretical lens. As the next section shows, accounts of food waste tend to focus on the acquisition of food for household consumption, and the impacts of packaging and food standards on the ways people appreciate food. It is suggested that such approaches present consumption as primarily a matter of individual action and food consumption as an almost entirely private domestic affair. Section three introduces theories of practice as a corrective to these over-emphases before exploring how we might begin to conceptualize eating practices and what light such understandings can shed on food waste. Section four briefly demonstrates the potential application of these ideas using a recent survey of eating patterns conducted in the UK ($N = 2784$) to show how the context of meal occasions holds significant explanatory value for predicting the production of surplus food. The chapter concludes by suggesting that a conceptual framework developed from theories of practice offers new avenues for research that circumvents the current myopic focus on particular aspects of food consumption, and the conceptual tendency to default to methodological individualism. This framework has three principal lines of enquiry: 1) to explore eating as a compound practice; 2) to examine the sequential organization of constituent activities from which compound eating practices are comprised; and 3) to take account of the inter-connections across the broader practices that make up everyday life.

Understanding food waste

In recent years, concerns about the origins and consequences of food waste have risen to prominence in debates about sustainability. A recent report by the Food and Agriculture Organization of the United Nations (FAO, 2011) estimated that globally, one-third of the food that is produced for consumption is wasted (1.3 billion[2] tonnes annually). Not surprisingly, the issue has gained political prominence, as illustrated by the European Parliament's resolution to reduce EU volumes of food waste by half. With respect to households, WRAP (2013) estimates that 19 per cent of all food and drink brought into UK homes in 2012 was thrown away, half of which was food and drink not used or prepared and one-third represented 'leftovers'. While the greater weight of policy attention has addressed the management of municipal waste streams (e.g. anaerobic digesters) or 'production-side' solutions in the supply chain, attention has steadily shifted toward food consumption as a major battleground for food waste reduction.

It is, of course, a gross simplification to divide attention to food waste between production-side waste management and consumer-side mismanagement of household food budgets. There are many forms of intermediation between the producers of food and those that consume and dispose of it. Retailers, food standards agencies, culinary experts (especially celebrity chefs and food critics) and restaurateurs, to mention a few, all have a part to play in connecting and mediating between the domestic consumer and the world of food production and distribution. Yet, all too often discourses about waste in food consumption present it as a personal domestic problem to be resolved either by voluntary lifestyle changes or through regulation to guide those food choices (e.g. information campaigns aimed at encouraging greater use of leftovers or the removal of food waste from general waste disposal services).

Evans *et al.* (2013) provide a comprehensive review of the history of food waste. Concentrating on food consumption, they demonstrate that concern with food waste is not as new a phenomenon as many social commentators would lead us to believe. Eighteenth-century cookery books and household management manuals were riddled with advice on how to re-use food and minimize waste, and were packed with commentaries extolling the moral virtues of thrift and frugality in the domestic kitchen and dining room. They argue, however, that the rise of the post-war global food regime characterized by its techno-industrial production systems, not only created an abundance of food but also made waste, temporarily, invisible.

This period of invisibility has only served to heighten contemporary anxieties regarding a food waste crisis in a time when environmental sustainability, food security and global recession have simultaneously come under the political spotlight.

Accounts of food waste in the sphere of consumption concentrate on two aspects: the accumulation of food items within homes that are never prepared for consumption and disposed of when they reach their use-by date; and, the leftovers from prepared meals that make their way, often indirectly, into the bin (Evans, 2012). Ethnographic studies of these two forms of food waste can be located within a number of familiar concerns about the conditions of contemporary forms of consumption (or consumer cultures) more generally.

Most prominent are accounts of food waste as being a problem of over-consumption. In some cases, over-consumption refers to volumes of food in general (often tied to concerns around obesity in an age of food abundance in affluent societies). In other cases, it refers to cultural politics about the profligacy of everyday consumption activities, or to the regulatory and market forces that produce packaging and standards that only serve to encourage and perpetuate the over-consumption of food items. To scholars in the sociology of consumption, the fears and anxieties concerning the socially corrosive effects of mass consumption are well-rehearsed – whether applied to consumer culture and well-being (Schor, 2010) or more generally to debates about sustainability (Jackson, 2009).

Of course, there are specificities of focus when this broader cultural anxiety is applied directly to studies of household food waste. Perhaps the two areas that have received greatest attention are food packaging and food standards or labelling. Located within what Alexander *et al.* (2013) term a 'post-humanist' theoretical orientation towards the study of food waste, the work of Hawkins (2013) explores the performativity of packaging on consumption. Drawing on, but significantly extending, earlier work on how packaging shapes and changes the meaning and use of products (e.g. Cochoy, 2007), Hawkins provides an account of packaging as actant. Using the case study of bottled water, she demonstrates how packaging – in this case plastic bottles – acts on consumers, affecting how water (in bottles) is understood (as fresh, healthy and so on), shaping how the bottles are held and used in day-to-day actions, and framing the spaces and places that the 'bottle can go'. Packaging is no longer simply presented as a problem in its own right, as was the case in earlier studies where food packaging was itself regarded as being 'wasteful or excessive' and something else to be disposed of. Rather, packaging has come to be held as performative, shaping the needs and wants of consumers, and leading to the burgeoning volume of food-related consumption.

Over-consumption not only focuses on volume but on the temporalities of food consumption. A significant problem, identified in numerous policy reports (e.g. WRAP, 2013), is that much food is disposed of without even being prepared – essentially moving from the supermarket shelf to the dustbin with a temporary pause while located in a household cupboard or fridge. This is over-consumption, but over-consumption that results from a temporal mismatch between the rates and frequencies of food acquisition and food consumption. Empirical (and policy) attention has focused on food standards and particularly on 'best before' and 'use by' dates that are located on all food products in the UK. Milne's (2013) account of how food labelling in the UK acted to codify cultural understandings and expectations of freshness and food quality are an excellent example. Best before dates, originally a method for shop workers to ensure stock rotation as opposed to informing consumers, have come to mark quality and a short shelf life indicative of freshness. Food standards do not encourage abundance with respect to the volume of food acquired, but they do affect cultural understandings of the satisfactory durations that food can be retained for.

A second, and related, area of empirical scrutiny has focused on thrift, frugality and re-use. In his earlier work on thrift, Evans (2011) explored how cultural conventions of frugality, including the re-use of material goods, can be understood across generational groups. Elderly respondents registered as displaying pro-environmental behaviours in terms of re-using goods and foods by making best use of leftovers, but demonstrated limited attitudinal enthusiasm for sustainability issues. The irony of those with anti-environmental values demonstrating pro-environmental behaviours is less significant than his argument that cultural skills and understandings are embedded in broader social processes – in this case related to age and generation. Food disposal, in this case, is as much a matter of cultural convention as it is of food abundance.

It is often assumed that thriftiness or frugality reduces food waste, as households are much more attentive to avoiding over-consumption and to ensuring re-use. Cappellini and Parsons (2013), however, demonstrate that thrift is a contingent and contextually specific cultural value. For example, thrifty meals are contextualized with respect to household economies of time, money and effort – the quick 'tasty' meal lending itself to re-use (or reheating) more so than a more elaborate 'slow' meal – spaghetti bolognese being a particular favourite of their respondents. The thrifty meal therefore needs to be understood in terms of the acquisition and assembling of particular types of meals (cheap and quick) and the expectations of the wider family (a tasty and filling meal). This is a highly skilled activity, which leads Cappellini and Parsons to conclude that thriftiness in the context of food consumption and disposal is a middle class (and possibly weekday) proclivity.

As the broader sociological literature on 'waste' emphasizes, domestic food management – apparently central to the production of consumer-driven food waste – is subject to a set of significant and competing moral tensions and cultural values. These include thrift, frugality and efficiency in provisioning and using food, but are not confined to them. The wider cultural context of eating shapes contemporary patterns of food consumption and domestic food management in ways that complicate a focus on household consumers as agents of change. From this brief survey of the literature on food consumption and food waste, six competing cultural demands or dynamics can be identified.

- *Food safety and health:* Strict guidelines about safe food maintenance and healthy eating encourage people to consider food in terms of possible health

risks (see, for example, Watson and Meah, 2013), potentially impeding efforts to use or re-use purchased food.
- *Variety and plenty*: Caution regarding over-consumption or overstocking of cupboards may be at odds with the cultural desire to have a ready stock of food available in the home for family and guests.
- *Care:* The responsibility to provide wholesome 'family meals' as a key process in producing family life may represent a cultural 'push' towards over-provisioning of healthy or comfort foods which are subsequently not used.
- *Convenience:* In contrast, juggling busy household schedules produces difficulties in the coordination of daily life (see, for example, Southerton, 2003) that shape the temporalities of food provisioning, eating and storage and may lead to unintended waste.
- *Economy:* Household budgeting processes, fundamentally subject to food costs and class, make economic concerns an overriding consideration in domestic food management.
- *Extravagance and indulgence:* Tensions rest between managing household budgets and providing for conspicuous, pleasurable and perhaps hedonistic, consumption occasions (see, for example, Wilk, 2014). Although tendencies towards extravagance and indulgence are often exaggerated in accounts of 'over-consumption', they are nonetheless in clear competition with the imperative to reduce or eliminate household-related food waste.

These competing demands are not presented here as being exhaustive, but to emphasize the point that food waste can only be fully grasped when located within the wider set of socio-cultural processes that affect contemporary eating.

Yet, with the exception of the studies reviewed above, dominant accounts continue to position the wasting of food as a discrete activity autonomous of the social and cultural contexts in which it is produced. This leads to approaches that identify 'barriers' to the reduction of food waste, whether those are related to labels or standards, packaging, or consumers' lack of skill and knowledge to manage domestic budgets. With respect to the latter, 'habits' and 'routines' that mindlessly lead to high levels of food waste are frequently presented as an obstacle to consumers adapting more efficient domestic food management systems. As argued elsewhere (e.g. Southerton, 2013), reducing repeated and socially embedded actions that involve consumption to the status of automated habits is a particularly problematic way of explaining everyday practices. No wonder authors such as Evans *et al.* (2013) criticize current policy discourses that locate the problem of food waste as being principally a matter of individual and private consumer resolution – to be addressed by consumers' taking responsibility for their wastefulness in the context of their homes.

Practices of eating

Following Warde's (2005) application of practice theories to the study of consumption – particularly the idiom that consumption occurs due to socially meaningful engagements with practices – attention to practice theories has grown, largely as a critique of the methodological individualism endemic in dominant accounts of consumer behaviour and its default explanations founded upon the 'portfolio model of action' (Warde and Southerton, 2012).[3] A key feature of these critiques are to proclaim practices the fundamental unit of social analysis (Shove et al., 2012) and to suggest that the recursive relationship between individual performances and social practices could provide a satisfactory basis for relating agency and structure, and exploring processes of social transformation.

The idea that practices are social and shared is commonly agreed, though explanations of the degrees to which practices are shared is contentious, depending on whether practices are considered as *entities* in themselves or merely similar *performances* repeated widely across time. When operationalized, practices are generally treated as configurations of recognizable, intelligible and describable elements which comprise their conditions of existence. While there is no single agreed typology of elements (compare options in Spaargaren, 2011 and Shove et al., 2012), some combination of material objects, practical know how and socially sanctioned objectives is deployed, these often in the context of socio-technical systems, social and economic institutions, and modes of spatial and temporal organization. Such elements and contexts both configure how practices are performed and make them identifiable to practitioners and non-practitioners alike. The relationship between practices and performances is recursive: practices configure performances, and practices are reproduced and stabilized, adapted and innovated, through performances. A critical theoretical question thus focuses on the tensions and dynamics between the reproduction (stability) of stable practices and adaptation (innovation) in the performance of practices that hinder or generate social change (McMeekin and Southerton, 2012).

One confounding problem for theories of practice is definitional: how might we identify a practice in order to study it, and how do we define its boundaries vis-à-vis any other collection of human activities? As Harvey et al. (2012) explain, such definitional questions are methodological in foundation. Definitions – or boundary drawing – of any practice are largely dependent on the form and focus of the research questions being asked. Enquiry into practice performances tends to define practices relatively tightly (e.g. showering, bathing, driving, cycling – to name but a few from recent studies – see Harvey et al. (2012) for a full discussion). Explorations of practices as entities tend to draw loose, more expansive, boundaries around 'bundles of activities' (e.g. practices related to achieving thermal comfort, mobility, working and mothering) as they seek to reveal the common elements that bind those activities together.

Warde (2013) attempts to address these definitional conundrums with respect to the practice of eating. His first step is to consider Schatzki's (1996) distinction between dispersed and integrative practices. The former refer to generic, usually tacit, practices that are dispersed across a range of activities. Cited examples include describing, following rules, explaining and imagining. Dispersed practices represent the basic 'know how' and 'understandings' for the appropriate performance of any activity. Integrative practices are 'the more complex practices found in and constitutive of particular domains of social life' (Schatzki, 1996: 98), with farming, cooking and business practices cited as examples. Integrative practices are thus socially organized and coordinated bundles of activities which produce practice performances that can be read and judged as correct and acceptable, and that are distinctly recognizable in their own right.

Warde argues that eating practices cannot adequately be defined as a dispersed practice and, at first glance, is:

> an activity [that] has many of the characteristics of an ideal typical integrated practice: it is a major activity of daily life; it is performed very often; it involves doing and saying; it has a large vocabulary devoted to it; much is written about it; it is a major activity of daily life; it involves understandings, procedures, specialized equipment and purposes; and it is instantly recognizable when encountered.
>
> *(Warde, 2013: 22)*

At a second glance, however, there are also a number of critical features of integrative practices missing. There are, at least in Britain, only generic shared understandings (of what it might mean to eat well), standards and rules that govern the activity.

Warde concludes that eating should be understood as a compound practice, which rests at the intersection of several integrative practices. Four integrative practices stand out, each with its own specific formalization in terms of rules, procedures and standards. These are: the supplying of food formalized in terms of understandings surrounding nutrition; cooking, formalized through codified instruction manuals (e.g. the recipe book); the organization of meal occasions, formalized through etiquette and manners; and, aesthetic judgements of taste that are formalized through gastronomy. These four integrative practices have developed at different speeds and according to different logics that have resulted in varying degrees of coordination and organization. The first two (supply and cooking) are more directly located within the sphere of production and have a greater degree of formal organization when contrasted with the latter two (meal occasions and aesthetic judgement) which belong to the sphere of consumption and are open to greater contestation.

Following Warde's line of reasoning, food disposal, which is intricately bound up with eating, could equally be understood as straddling its four integrative practices. The focus of much existing research can, retrospectively, be positioned within such a description, with an emphasis on the integrative practices of supply (packaging, food standards and labels) and cooking (over-preparation, convenience foods). With the exception of the studies discussed by scholars such as Evans (2012), Watson and Meah (2013) and Cappellini and Parsons (2013), much less attention with respect to food disposal has been focused on the social organization of meal occasions and aesthetic judgements of taste. Yet, accounts which highlight the tensions between moral orders are, in their own way, grappling with the intersection, contestation and incommensurability of the integrative practices from which eating (and food disposal) as a compound practice is constructed.

The major intellectual advantage of applying Warde's understanding of eating as a compound practice is that it provides a conceptual framework that offers the opportunity for systematic analysis of food disposal as a constitutive element of a range of integrative practices. It 'holds together' the relationship between food acquisition, preparation, meal occasions and cultural tastes without reducing the performance of food disposal to any one. In doing so, it provides an approach, which emphasizes the emergent properties of contemporary eating practices, and how those properties relate to each other, while offering the opportunity to analyse and indicate which integrative practice (or practices) is more or less significant for explaining food waste.

Understanding the production of surplus food in the context of eating

Operationalizing Warde's conceptual framework to analyse eating and food disposal requires a microanalysis of the constituent activities from which the practice is comprised in order to understand the relationships between those activities. Such a deconstruction is common in time-diary studies; a methodological tool that facilitates analysis of how the temporal (and spatial) coordination and sequencing of activities render blocks of shared activities recognizable as practices. This approach was developed in previous research that employed time-diary data to analyse the changing form of eating practices in the UK over a 25-year period. In this study, we de-constructed the practice of eating into a set of constituent activities: food preparation, eating at home, eating at the homes of others and eating out (see Cheng *et al.*, 2007), which were the only discrete and relevant activities recoverable from that dataset. In doing so, we were able to chart the relationships between each activity in configuring contemporary patterns of food consumption in the UK, showing that, contrary to popular fears built around a focus on one constituent eating activity (eating at home), the family meal has not disappeared but has shifted outside the domestic setting.

The critical point is that eating practices (and the waste generated) are best operationalized for purposes of analysis as complexes or sequences of activities. From this perspective, the central research question is whether different formations of constitutive activities encourage or permit more or less food waste, as well as what forms of cultural understandings and conventions underpin the contexts in which those sequences of activities are performed.

Our contention would be that, as with analysis of social change in eating practices, the analysis of food waste must acknowledge the context of the constitutive activities in which eating patterns are located. By way of example, we draw on contemporary empirical data from the UK, which includes information about eating events in context and the production of food waste in the form of leftovers to explore the conditions of eating events in which 1) food surplus or leftovers are generated (specifically, food which is produced or served at the meal event but is not eaten during it); and, 2) briefly, where food 'waste' is noted (where the food is explicitly reported as thrown away).

An online survey was administered to panel members of a major UK supermarket loyalty card scheme in September 2012. Respondents were asked to describe their eating activities through a time-diary formatted survey. Socioeconomic and demographic variables were obtained. The response rate was 45 per cent and the achieved sample size was 2,784. As self-selecting members of a consumer panel, the sample is not representative of the general population, and older, more affluent and better educated respondents are somewhat over-represented. Nevertheless, the data is sufficiently copious, diverse and recent to be worth careful attention for our purposes.

Respondents were asked to report on their eating activities on the weekday prior to filling in the questionnaire and one day during the previous weekend. Each was asked to detail up to seven eating events for each day reported upon, providing a description of the food in their own words and answering subsequent questions about when, for how long, with whom and where it was eaten, as well as its origin and preparation. In addition, the survey asked respondents to estimate if, and how much, of the food was left over after the eating event, with a final question about the use, storage or disposal of this surplus food. As such, this survey data provides information about one potential route through which food disposal is said to occur: the production or serving of surplus food which is subsequently unused. The following analysis explores the context in which surplus prepared food or leftovers are generated, in approaching two of the integrative practices from which eating is composed: the preparation of food and the organization of the meal occasion. We operationalize this conceptual frame by analysing relationships between the following constituent activities of eating practices: food provisioning; preparation; the meal occasion (i.e. the length of the episode, whether and what type of company was present, and the meal time); surplus reported; and the disposal of any surplus food. For reasons of brevity and because they are generally understood as the 'main meals' of the day (see Marshall 2005), we confine our analysis to the first or only meals, for each respondent, taken between midnight and 12 pm, known henceforth as morning meals or breakfasts, those taken between 12 and 5pm, known henceforth as afternoon meals or 'lunches', and those taken between 5pm and midnight, as evening meals or 'dinners'.[4]

The production of leftovers does not necessarily or directly lead to food waste. Surplus food is prepared deliberately in some households and stored for use in subsequent eating events, although it is debatable whether this food would be considered and reported as leftovers. A variety of 'routes' exist for the disposal of unused household food, with forms of re-use or recycling possible by deploying fridges, freezers and reheating technologies; feeding leftovers to household pets; and composting. Of the occasions on which leftovers were registered by respondents, Figure 8.1 reveals that around one-quarter of the leftovers prepared or served at eating occasions was simply thrown away, with a further one-third eaten

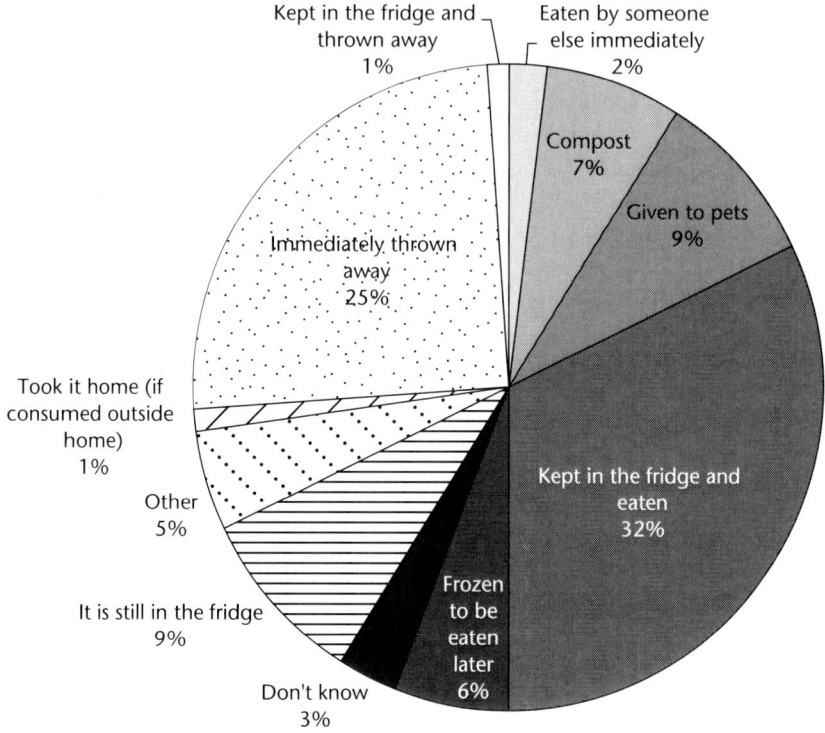

FIGURE 8.1 The destinations of surplus food produced at afternoon and evening meals
Survey question wording: 'What happened to the leftovers?'

between the point of preparation and the survey's application, with 16 per cent redirected to compost or pets and 15 per cent stored for potential re-use or for eventual disposal. The following analysis concentrates mainly on the category of leftovers *per se*, in part because we would estimate, following Evans (2012), that significant proportions of the surplus food from meals is never re-used, but also that the use of composts and pets could be considered a form of mitigated waste. Additional attention is paid to the 25 per cent of the surplus that was reported as explicitly 'thrown away' and that can be defined with some certainty as food waste.

Table 8.1 reports on the relationship between our constituent activities of eating practices (provisioning, preparation and the contexts of meal occasions – represented by variations of meal duration, commensality and timings) and the respondents' reports of leftovers at meal occasions and their disposal. It shows the percentages of the first afternoon and evening eating events coded by each respondent as 'main meals', thus pertaining to what are commonly described as lunch and dinner in which food surplus, and food waste, is reported. Overall, some leftovers are reported at 22 per cent of these main meals, from which roughly a quarter are thrown away (5 per cent of the meals reported).

TABLE 8.1 Characteristics of afternoon and evening meals at which surplus food and food waste is reported

	% meals with leftovers	% meals with leftovers thrown away	N
Food provision			
Home	20.4	4.6	7194
Friends/family	39.0	6.4★	172
Restaurant/takeaway/shop	25.7	7.9★	1234
Other	22.3	6.8★	530
Combination of origins	21.7	5.8★	138
Food preparation			
Not cooked/ready to eat	5.5	1.4	1696
Basic preparation#	5.3	2.3	533
Complex preparation#	30.4	6.6	4593
Duration			
Under 20 minutes	13.0	3.7	5264
20–39 minutes	30.6	7.0	3282
Over 40 minutes	48.2	8.9	596
Commensality			
Alone	7.8	2.3	2214
Household members	26.3	6.4	5456
Other companions/combinations	24.1	5.2	1549
Which meal			
Weekday lunch	6.9	2.5	2355
Weekday dinner	25.6	6.9	2420
Weekend lunch	21.9	4.7	2142
Weekend dinner	31.8	6.7	2351
All	21.6	5.2	9268

Survey question wording: 'How much was left over?' Answer response categories: Nothing; Less than 10%; Between 10%–30%; Between 30–50%; More than 50%.

Basic preparation entails use of a toaster and/or microwave; complex preparation means use of a hob and/or cooker.

★ Figures for foods reported as thrown away are also included here, but should be interpreted with caution, given that those eating cannot usually accurately ascertain the destination of leftovers produced outside the home.

Different provisioning sources of the meals eaten (whether they were home prepared or came from restaurants, takeaways, directly from shops, friends or family, workplaces or some combination of these sources) produce very different proportions of meals where there is food surplus. Meals prepared by friends or family, as in the case of dinner parties, are twice as likely to involve leftovers as are those originating from home. Focusing on meals prepared at home suggests that more lengthy and complicated preparation processes also produce leftovers more often: 6 per cent of home-prepared afternoon and evening meals that involved no preparation produced leftovers, compared with 30 per cent of meals that involved cooking. There is a strong correlation between the reporting of food surplus, with meal origin or form of preparation.

Meals of longer durations, evening meals and meals where other people were present are also much more likely to produce leftovers. Meals eaten with others were over three times more likely to involve leftovers than those eaten alone. Meals of long durations (duration representing an indicator of conviviality) resulted in higher levels of both leftovers and waste, and lunchtime meals were responsible for much less surplus and waste than were dinnertime meals, although weekend lunches generated much greater volumes of surplus food and waste than weekday lunches.

A basic description of the data demonstrates apparent associations between food provisioning, preparation and meal occasions – with complex and more convivial meal occasions appearing most likely to generate surplus food and subsequent waste. To further explore these relationships a logistic regression model, which analyses the relationship of each factor while controlling the others, using the dependent variable of incidence of leftovers (*vs* no leftovers), was developed.

The model (reproduced in Table 8.2) shows that the associations visible in the percentage distribution are strong and highly statistically significant, and the model is estimated as accounting for 22 per cent of the variation. Meals from home that are cooked using the hob or oven are nearly five times more likely to produce leftovers than those which were merely assembled without cooking. The apparent effect of meals originating outside the home remains strong, with these 3.6 times more likely to produce leftovers than those meals that were ready-to-eat and from the home. Meals lasting over 40 minutes are over three times more likely to produce leftovers than those eaten in under 20 minutes, and the presence of companions – particularly those from outside the household when other factors are controlled for – had a similar impact. The effect of the meal occasion and timing itself also remains significant, with evening and weekend meals between two and three times more likely to create leftovers than the reference category of weekday lunches.

Running a similar logistic regression for wasted food at meals – the leftovers we know from the survey to have been wasted – shows several interesting differences, although the direction of relationships with reference categories remained consistent with that of Table 8.2. Meals prepared elsewhere are reported as more wasteful than meals cooked in the home, although home cooking produced more leftovers. It was unclear from the data whether respondents assumed that leftovers produced in restaurants are automatically wasted. Longer meal durations are slightly less strongly associated with the production of wasted food than they are with leftovers. Meals taking longer than 20 minutes in length are only about one and a half times more likely to involve food being thrown away when compared with meals under 20 minutes, whereas leftovers were two or three times more likely to occur at longer meals. The presence of others is not as strongly associated with food waste as it is with leftovers, and the presence of non-household members appears to make the leftovers slightly more likely to be used than if the meal is shared only with household

TABLE 8.2 Binary logistic regression of food surplus at afternoon and evening meals

	β	S.E.	Exp (β)
Food provision and preparation#			
Ready to eat (home) (ref)			
Basic preparation (home)	−0.039	0.229	0.962
Complex preparation (home)	1.554★★★	0.118	4.731
Prepared outside home	1.289★★★	0.130	3.631
Home prepared, eaten elsewhere	0.358★★	0.207	1.430
Duration			
Under 20 minutes (ref)			
20–39 minutes	0.674★★★	0.059	1.962
Over 40 minutes	1.189★★★	0.100	3.285
Commensality			
Alone (ref)			
Household members	0.796★★★	0.090	2.217
Other companions/combinations	0.998★★★	0.112	2.713
Which meal			
Weekday lunch (ref)			
Weekday dinner	0.802★★★	0.107	2.231
Weekend lunch	0.917★★★	0.107	2.501
Weekend dinner	1.100★★★	0.105	3.005
Constant	−4.363★★★	0.150	0.013
N	9092		
Nagelkerke r2	0.221		

Significance ★★★ $P < .001$, ★★ $P < .05$, ★ $P < .10$.

Food provision and preparation necessarily collapses two factors as only those meals originating and eaten at home had data on preparation.

members, despite leftovers being more common with the former than the latter. Possibly leftovers from meals eaten with others tend to be those considered worth saving. Finally, although leftovers are more commonly produced at weekend meals, when examining that which is thrown away, weekday evening meals appear to be slightly more likely to incur actual waste than weekend meals. Associations remained statistically significant to the 1 or 5 per cent level with the exception, as before, of comparing simple or 'warmed' preparation with those meals reported as ready to eat.

Conclusions

The core argument of this chapter is that if we are to fully understand food waste then attention needs to be directed at the practices in which food disposal is embedded. This re-directs theoretical and empirical attention away from myopic analyses of particular aspects of food consumption towards a situating of food waste within the fabric and processes of everyday practices. This offers possibilities for thinking through, and potentially re-orientating, broader systems of food consumption and the organization of the compound practice of eating in ways that could have ameliorating effects on food waste.

A (necessarily) brief demonstration of the utility of analysing the relationship between constituent activities of eating and the production of leftovers and food waste offers support for thinking about food waste in the context of eating practices more generally. The provision and preparation of meals are critical determinants of whether meal surpluses, and food waste, are produced. The processes of cooking and meal provisioning from other sources suggest that more socially significant meals are subject to practical and discursive pressures towards plenty and abundance which cannot easily be squared with an ethic of thrift. The impression is borne out in the analysis of the meal occasions themselves, where more leisurely meals, the presence of others, particularly from outside the household, and particular occasions with their own cultural conventions, all appear associated with the production of surplus and waste.

Our empirical conclusions are preliminary and four-fold. First, our data demonstrate that meal occasions, and the circumstances framing and enabling them, have important implications for the production of food waste, recommending further in-depth analysis of the contexts in which food is provisioned, prepared, eaten and disposed of. Second, the circumstances in which leftovers are produced appear to underline the significance of moral and cultural dynamics around the display of care, food diversity and extravagance in performances of home-making and commensality. Third, the importance of duration and of the variables around preparation, companionship and provisioning also suggest that the practical temporal coordination of the constituent activities of eating practices have a strong conditioning effect on whether surplus and waste are produced at eating occasions. Provisioning and food preparation for meal occasions that involve commensality (which itself requires the temporal coordination of eating practitioners) relate to longer meal durations and greater generation of surplus food. Interestingly, despite controlling for these factors, evening and weekend meals still produce more leftovers and food waste, suggesting that there are significant social or symbolic aspects of meal occasions and their timings that are not captured in our analysis. Finally, while not recoverable from the data presented here, our findings hint at the relationships and differences in eating with others in and outside the home. In different contexts, the cultural conventions around what is considered a satisfactory or 'good meal' – in terms of content, company, copiousness and aesthetics – are likely to vary considerably.

Beyond these empirical observations, the approach presented in this chapter sought to operationalize a practice theoretical lens for analysing food waste. In taking this approach, we highlight three more general observations about the value of practice theories for shedding new light on substantive problems associated with consumption.

The first, following Warde (2013), is the utility of conceptualizing complex practices such as eating as compounds of integrative practices – in this case of: food supply; cooking; organization of meal occasions; and, aesthetic judgements of taste. It is the formalization of these integrative practices that configure cultural understandings and conventions, and we suggest that it is the contested relationships between these integrative practices that underpin the kinds of moral-cultural tensions identifiable from existing studies of food waste.

The second calls for detailed analysis of the contextualized sequencing of the constituent activities from which compound practices such as eating are configured. As a means of empirically operationalizing a conceptual framework, we utilized time-diary techniques to deconstruct eating practices and analyse their constituent activities. In doing so, we reveal that sequences of provision, preparation, meal occasions and disposal vary across contexts, and those contexts are likely to affect the meanings and social organization of each constituent activity. Food waste, like food consumption more generally, is embedded in sequences of activity that can be described as routine performances, but are explained through analysis of the temporal, spatial and interactional (in this case represented through forms of commensality) ordering of those activity sequences.

Finally, and less visible in data presented here, is the importance of accounting for the relationships between multiple compound practices. It is likely that the patterns of eating presented here – such as the differences between weekday and weekend lunches – are strongly influenced by practices of working and mobility. Further and more sustained analyses of the inter-connections between these three areas are needed to properly understand and shape change in the practices themselves.

Notes

1 The research presented in this chapter was co-funded by the Economic and Social Research Council (ES/L00514X/1) and Tesco plc, with additional support from the Sustainable Consumption Institute, the University of Manchester. We are particularly grateful to Dunnhumby for their assistance with data collection and preparation. We would also like to thank Alan Warde, David Evans and participants of the '*Reduce, reuse, recycle – environmental, economic and social challenges from a consumer's perspective*' Workshop (University of Borås, Nov, 2013) for their comments and reviews.
2 Throughout this chapter, billion means 10^9 or a thousand million.
3 Whitford (2002: 325) defines 'the portfolio model of the actor' as understanding action in the following terms: 'individuals carry a relatively stable and pre-existing set of beliefs and desires from context to context. Given the situation, they select from this portfolio "those elements that seem relevant and [use] them to decide on a course of action".'
4 Fewer than 4 per cent of morning meals involved any leftovers, of which around a quarter was reported as thrown away.

References

Alexander, C., Gregson, N. and Gille, Z. (2013), 'Food waste', in A. Murcott, W. Belasco and P. Jackson (eds), *The Handbook of Food Research*, London: Bloomsbury.

Cappellini, B. and Parsons, E. (2013), 'Practising thrift at dinnertime: Mealtime leftovers, sacrifice and family membership', in D. Evans, H. Campbell and A. Murcott (eds), *Waste Matters: New perspectives on food and society*, London: Blackwell, pp. 121–134.

Cheng, S-L., Olsen, W., Southerton, D. and Warde, A. (2007), 'The changing practice of eating: Evidence from UK time diaries, 1975 and 2000', *British Journal of Sociology*, 58(1): 39–61.

Cochoy, F., (2007), 'A sociology of market-things: On tending the garden of choices in mass retailing', in M. Callon, Y. Milio and F. Muniesa (eds), *Market Devices*, London: Blackwell, pp. 109–129.

Evans, D. (2011), 'Thrifty, green or frugal: reflections on sustainable consumption in a changing economic climate', *Geoforum*, 42(5): 550–557.

Evans, D. (2012), 'Binning, gifting and recovery: The conduits of disposal in household food consumption', *Environment and Planning D: Society and Space*, 30(6): 1123–1137.

Evans, D., Campbell, H. and Murcott, A. (2013), 'A brief pre-history of food waste and the social sciences', in D. Evans, H. Campbell, and A. Murcott (eds), *Waste Matters: New perspectives on food and society*, London: Blackwell, pp. 5–26.

FAO (2011), *Global Food Losses and Food Waste: Extent, causes and prevention*. Food and Agriculture Organization of the United Nations, Rome.

Harvey, M., McMeekin, A., Shove, E., Southerton, D. and Walker, G. (2012), *Researching Social Practice and Sustainability: Puzzles and challenges*, SPRG Discussion Paper, No. 2. Available online at http://www.sprg.ac.uk/uploads/practices-and-methodological-challenges.pdf (accessed 13 August 2014).

Hawkins, G. (2013), 'The performativity of food packaging: Market devices, waste crisis and recycling', in D. Evans, H. Campbell and A. Murcott (eds), *Waste Matters: New perspectives on food and society*, London: Blackwell, pp. 66–83.

Jackson, T. (2009), *Prosperity Without Growth: Economics for a finite planet*, London: Earthscan.

Marshall, D., (2005), 'Food as ritual, routine or convention', *Consumption Markets & Culture*, 8(1): 69–85.

McMeekin, A. and Southerton, D. (2012), 'Sustainability transitions and final consumption: Practices and socio-technical systems', *Technology Analysis & Strategic Management*, 24(4): 345–361.

Milne, R. (2013), 'Arbiters of waste: Date labels, the consumer and knowing good, safe food', in D. Evans, H. Campbell and A. Murcott (eds), *Waste Matters: New perspectives on food and society*, London: Blackwell, pp. 84–101.

Schatzki, T. (1996), *Social Practices: A Wittgensteinian approach to human activity and the social*, Cambridge: Cambridge University Press.

Schor, J. (2010), *Plenitude: The new economics of true wealth*, New York: The Penguin Press.

Shove, E., Pantzar, M. and Watson, M. (2012), *The Dynamics of Social Practice*, London: Sage.

Southerton, D. (2003), '"Squeezing time": Allocating practices, co-ordinating networks and scheduling society', *Time & Society*, 12(1): 5–25.

Southerton, D., (2013), 'Temporal rhythms, habits and routines: From consumer behaviour to the temporal ordering of practices', *Time and Society*, 22(3): 335–355.

Spaargaren, G. (2011), 'Theories of practice: Agency, technology and culture. Exploring the relevance of practice theories for the governance of sustainable consumption in the new world-order', *Global Environmental Change*, 21: 813–822.

Warde, A. (2005), 'Consumption and theories of practice', *Journal of Consumer Culture*, 5(2): 131–154.

Warde, A. (2013), 'What sort of a practice is eating', in E. Shove and N. Spurling (eds), *Sustainable Practices: Social theory and climate change*, London: Routledge, pp. 17–30.
Warde, A. and Southerton, D. (2012), 'Social sciences and sustainability', in A. Warde and D. Southerton (eds) *The Habits of Consumption*, Helsinki: Open Access Book Series of the Helsinki Collegium of Advanced Studies.
Watson, M. and Meah, A. (2013), 'Food, waste and safety: Negotiating conflicting social anxieties into the practices of domestic provisioning', in D. Evans, H. Campbell and A. Murcott (eds), *Waste Matters: New perspectives on food and society*, London: Blackwell, pp. 102–120.
Whitford, J. (2002), 'Pragmatism and the untenable dualism of means and ends: Why rational choice theory does not deserve paradigmatic privilege', *Theory & Society*, 31: 325–363.
Wilk, R. (2014), 'Consumer cultures past, present and future', in D. Southerton and A. Ulph (eds), *Multi-Disciplinary Approaches to Sustainable Consumption*, Oxford: Oxford University Press.
WRAP (2013) *Household Food and Drink Waste in the United Kingdom 2012*, http://www.wrap.org.uk/sites/files/wrap/hhfdw-2012-main.pdf (accessed 7 March 2014).

9
ENVIRONMENTAL CONSUMER SOCIALIZATION AMONG GENERATIONS SWING AND Y

A study of clothing consumption

Karin M. Ekström, Daniel Hjelmgren and Nicklas Salomonson

Introduction

An accelerated pace of consumption with an abundance of goods and services to choose from as well as an increased significance of symbolic consumption have characterized recent decades (e.g. Ekström 2013). This is particularly noticeable for consumption of clothing. Fast fashion and low prices have, during the last decades, contributed to a significant increase in consumption in several countries. In Great Britain, the volume of sold clothing increased by 60 per cent between 1995 and 2005 (Morley *et al.* 2006). In Sweden, private consumption of clothing and shoes increased by 53 per cent between 1999 and 2009 (Roos 2010).

Parallel to an increase of purchases, we are also witnessing an increase in the waste of clothing. Assessments by the EPA Office of Solid Waste in the United States indicate that Americans throw away more than 30 kg of clothing and textiles per person per year (Claudio 2007). In a similar manner, an average consumer in the United Kingdom throws away 30 kg of clothing and textiles each year (Allwood *et al.* 2006). A pick-analysis study indicates that Swedes throw away about 8 kg textiles per person per year (Carlsson *et al.* 2011). This is a waste of resources and it would be more environmentally sound to reuse or recycle clothing (e.g. Ekström *et al.* 2012). Production of clothing calls for a large amount of natural resources. For example, one kilogram of cotton requires about 7,000 to 29,000 litres of water and 0.3 to 1 kg of oil (Fletcher 2008). In addition, chemical pesticides are often used in the production.

Different generations experience consumption differently, depending on prevailing societal norms and values, economic factors and experiences of consumer culture. Consumers who have faced economic downturns are likely to have a different attitude compared to consumers who have been brought up during more prosperous times. In this chapter, we are interested in finding out how Generation Swing (born 1930–1945) and Generation Y (born 1976–1994) experience and approach consumption of clothing. Generation Swing has experienced difficult economic times during their upbringing, and lately an emergent consumer culture. Generation Y has experienced an accelerated pace of consumption and is considered the most consumption-oriented generation of all time due to an abundance and availability of products and services (Sullivan and Heitmeyer 2008). They are also claimed to be more socially and environmentally conscious than earlier generations (Sheahan 2005). The chapter is based on focus group interviews in Sweden. Before the method and results are presented, we will briefly discuss the theoretical basis for our study.

Generations Swing and Y

During their upbringing in Sweden, Generation Swing (born 1930–1945) and Generation Y (born 1976–1994) have encountered different societal norms and values that are likely to influence their consumption patterns. Consumption during the 1930s was influenced by the worldwide recession, and the welfare state was in its formative phase (Mattsson 2010). The beginning of the Second World War in 1939 led to a lack of goods. People were encouraged to reuse and recycle by caring for clothing (e.g. air clothing rather than washing), patching and re-sewing. Things were often repaired rather than being replaced. Children learnt how to mend and patch clothing in school and also at home. Clothing that was no longer usable was recycled into mattresses and paper. Apart from practicality during this period, there was a strong emphasis on cleanliness (e.g. Nordström 1938).

The Yuppie (young urban professional) era during the 1980s placed a stronger focus on brands. The collectivist values during previous decades were replaced with a stronger focus on individual preferences. Lindgren *et al.* (2005) argued that Swedes who grew up during the late 1980s are more concerned about doing things that create value for themselves and have a stronger need for expressing their own identity, for example by dressing more individually. They like to shop, and when asking them what they would do if they could spend double as much as they do today, the answer is: purchase more clothing and shoes (Lindgren *et al.* 2005). People who were born during the 1980s are often aware of how they through consumption can influence their identity (Parment 2008). Clothing plays an important role in displaying identity.

Consumer socialization and consumption of clothing

Consumer socialization helps us to The understand of how consumers relate to shifting consumption norms and ideologies and to cultural as well as technological changes (Ekström 2006). This chapter contributes to the understanding of how consumers orient themselves towards consumption of clothing, including purchases, use, reuse, recycling and disposal. Ward (1974: 2) defined consumer socialization as 'the process by which young people acquire skills, knowledge, and attitudes relevant to their functioning as consumers in the market place'. Even though most research on consumer socialization has focused on children or adolescents (e.g. Churchill and Moschis 1979; Moschis 1985; Carlson and Grossbart 1988), it is a life-long process (e.g. Brim 1966; Ekström 2006; Ward 1974). The influence of different socialization agents (family, friends, colleagues, school, media etc.) differs, however, at different periods in life.

Consumer socialization is often studied in the context of consumer learning, but learning can occur in many ways; directly by explicitly being told something or indirectly by observation or modelling. Socialization is often more subtle than purposeful (Ward 1974). Tallman et al. (1983: 23) wrote that in most cases 'socialization takes place indirectly as an unintended, or at least, implicit consequence of an ongoing relationship'. A person may first become socialized and then become the socialization agent (Ekström 2006). Children may, for example, learn things in school and then share this with their parents (Ekström 1995; Moschis 1976). Consumer socialization throughout life may also result in a person rebelling against behaviour that has been taught during childhood (Ekström 2006). Berger and Luckmann (1967: 150) defined primary socialization as 'the first socialization an individual undergoes in childhood, through which he becomes a member of society'. Secondary socialization is defined as 'any subsequent process that inducts an already socialized individual into new sectors of the objective world of his society' (ibid.). In a continuously changing society, secondary socialization can be expected to play an important role (Ekström 2010). For example, adult consumers learn how to behave in a more environmentally friendly way in a society that encounters environmental problems. There are studies on environmental consumer socialization among children (e.g. Grönhöj 2007), but there is a need for more studies on environmental consumer socialization of adults. Researchers also need to study how early socialization affects later socialization and whether and when early socialization hinders or facilitates later socialization (Ekström 2006).

Socialization entails conformity to societal requirements or norms (Ekström 2006). Moschis (1987: 23) wrote: 'from a societal perspective, an individual can be said to be socialized when he or she has learnt to think and feel according to society's expectations'. Socialization is closely tied to being a member of society (Ekström 2006). However, the answer to whose responsibility it is to teach consumers to function better as consumers is likely to differ depending on market ideology (Ekström 2006).

This chapter focuses on Sweden, where consumption of clothing has increased significantly during the last decades. In total, 62 per cent of the Swedish population purchased clothing at least once a quarter during 2012 (Ekström 2013). Even though it was predominantly younger or middle-aged consumers (80 per cent of consumers between 16 and 29 years of age, 66 per cent of consumers between 30 and 64 years of age), older people are also active consumers (43 per cent of consumers between 65 and 85 years of age). Another study (Ekström *et al.* 2012) found that the most frequent purchasers of clothing were younger or young middle-aged women. Economic (disposable income, opportunities to borrow money and low prices), social (symbolic consumption) and cultural factors (interest in fashion and fast fashion) have contributed to this development (Ekström and Salomonson 2012; 2014). The reasons for buying clothes can be to find something new or to replace something, either because of novelty or that something is worn out.

Parallel to increased consumption, there is also an increase in the waste of clothing. A Swedish study shows that 21 per cent of people throw away clothing because they are tired of it, and that young adolescents (16–19 years old) are more likely than others to think this way (Ungerth 2011). Other reasons for disposal of clothing could be that collection bins are not easily accessible or that consumers do not know which charitable organizations to trust (Ekström *et al.* 2012). Furthermore, consumers may not be aware of how their consumption of clothing affects the environment. In Sweden, consumers during the last decades have been consumer socialized to recycle different parts of waste. For example, in 2011, 92 per cent of glass, 74 per cent of paper packaging and 68 per cent of metal were recycled (FTI 2013). One reason for this is that a national system has been developed and that over time consumers have learnt to recycle. Currently, there is no national system for reuse/recycling of clothing and textiles in Sweden. Also, consumers are not always aware that torn clothing can be recycled. Recycling of textiles was common up until a few decades ago, but is today only used to some degree in the production of new clothing. For example, Patagonia makes fleece clothing from recycled polyester. Recycled material is also used as rags, soundproofing in cars, and stuffing in mattresses and furniture (Morley *et al.* 2009).

Other reasons for disposal of clothing could be that repair, patching and mending have become less common, in particular if it is more costly than to buy something new. Such skills have commonly not been prioritized, neither in school nor at home, during recent decades. However, many of today's young consumers have been consumer socialized to behave in an environmentally friendly way in kindergarten and at school. This was not at all prioritized when older consumers were brought up, but instead they had a frugal lifestyle that might resemble environment consciousness (Ekström *et al.* 2012). It is foremost throughout their adult life that older consumers have learnt about environmental friendly consumption behaviour.

Method

Six focus group interviews were conducted to get a more in-depth understanding of Generations Swing and Y's disposition to consumption of clothing. These were two groups of non-environmentalists representing Generation Y (n = 6 and 10), two groups of non-environmentalists representing Generation Swing (n = 7 and 10), one group of environmentalists representing Generation Y (n = 9) and one group of environmentalists representing Generation Swing (n = 9). All participants were middle-class and native-born Swedes. Homogeneous groups, in terms of cohort and engagement in sustainability issues, enabled comparisons in terms of orientations towards consumption of clothing regarding purchases, use and disposal. The participants were recruited by contacting a book circle (senior citizens), an environmental organization and a university (students and employees). All participants were offered an incentive (a towel or a coffee cup with the university logo) for their participation.

An interview guide was developed, and each focus group interview lasted about two hours. Two of the focus group interviews were conducted in a meeting room at the university, two at the homes of the participating senior citizens and two at the office of an environmental organization. Refreshments were provided as a way to relax the atmosphere, and thus stimulate interaction (e.g. McDaniel and Bach 1996). Each focus group was led by one or two researchers who acted as moderators during the meetings.

Results and analysis

The results and analysis regarding shopping patterns, shopping for second-hand, care of clothing, disposal of clothing, giving to charity and trading clothing are presented for both Generations Y and Swing, but separately for the non-environmentalists and environmentalists.

Shopping patterns

Non-environmentalists

The study shows that members of Generation Y who were non-environmentalists shop more often and spend more time shopping compared to Generation Swing. Generation Y often shops without knowing exactly what they are looking for. At other times, they plan their purchases carefully and visit a large number of stores. They enjoy the search process, probably as a result of enjoying shopping. Generation Y shops for new clothing or replacement of clothing they are tired of, while Generation Swing shops mainly for replacement of worn-out clothing. Since fashion changes constantly, Generation Swing expresses difficulty in finding exactly what they are looking for. For example, if a turtleneck needs to be replaced, it may be out of fashion and hard to find:

> Suddenly, you have no turtleneck anymore and you need to purchase a new one. It can be difficult if it is no longer in fashion.
>
> *(Margareta, 73)*

Generation Y prefers to purchase clothing (primarily fast fashion) in department stores and on the Internet, while Generation Swing primarily visits boutiques and outlets.

Different experiences during primary socialization may explain the differences between the generations. Generation Swing was brought up and consumer socialized at a time when shopping behaviour was more restricted. Generation Y, on the other hand, has experienced an emerging consumer culture involving fast fashion, cheaper prices and the possibility to borrow money to consume.

Another difference between the generations is that Y prioritizes price, style and fashion while Swing focuses on quality and durability when purchasing clothing. Preference for quality and durability means that Swing often buys classic design:

> I like to purchase classical design. I do not like to buy and throw it away shortly after. It is not for me. I want to purchase a garment that I enjoy, that I can wear for a long time if it fits me well, has good quality and I feel that I like it.
>
> *(Britta, 69)*

Preference for price, style and fashion results in Generation Y buying fast fashion even when they know it is not the right thing to do:

> You always forgive the giants .... The reason is that it's cheap. You think it's really awful that [name of retailer] uses child labour, thinking I'll never shop there. But then you go there again, you need a sweater and they sell it for 149 kronor. The alternative is to pay 500 kronor. I choose to pay 149 kronor.
>
> *(Evelina, 24)*

Again, primary socialization may explain the differences between the generations. As children, Generation Swing was socialized to buy clothing according to their functional needs. In contrast, during their upbringing, Generation Y experienced that clothing is used to express identity. Generation Y has also experienced a higher availability of cheap fashionable clothing during their upbringing. This can explain why style, fashion and more hedonic motives are prioritized in their consumption of clothing. The significance of fast fashion during recent decades (e.g. Byun and Sternqvist 2008) needs to be recognized when understanding the different generations' preferences.

Environmentalists

Members of Y and Swing who are environmentally engaged are trying to limit their consumption of clothing. Most of the clothes they purchase are second-hand. Generation Swing seldom buys new clothing and when they do, it is because something is damaged or because it does not fit any more. For Generation Y, shopping also involves renewing the wardrobe. One explanation for this difference could be that Generation Y has been brought up during decades when symbolic consumption has increased significantly (e.g. Ekström 2013). However, both generations try to minimize visits to retail stores because of environmental concern. Some of them also said that they get overwhelmed and try to cope with that by planning their purchase:

> I feel bad when I go into stores. I almost get giddy and think – no, I can't cope with it. It is enough that I'm standing on the threshold [of the store]. It can sometimes also be like this when I look for second-hand. So when I shop, it is often quite planned.
>
> *(Maria, 28)*

Another strategy used by Generation Swing is to always visit the same store, often a small one with good service. This makes shopping less demanding. When buying new clothing, Swing prefers more expensive stores with higher levels of service and quality of clothing. Generation Y mostly buys new clothing in low price stores, since they have found that such stores also can have good quality:

> Sometimes I find really nice 100% wool sweaters at H&M and think – my god what happened. In other cases, it has been enough to wash the garment once and then it has turned and twisted beyond recognition.
>
> *(Elin, 28)*

Both generations find quality and durability to be important when choosing clothing. It is more important than fashion. The fact that clothing needs to last longer is also a reason for buying second-hand:

> The quality of the clothes used to be much better, that's a reason why I often buy second-hand.
>
> *(Emma, 28)*

Both generations have learnt about the benefits of acting environmentally soundly during secondary socialization, in other words during their adulthood.

Shopping for second-hand

Non-environmentalists

Both generations are relatively reluctant to buy second-hand clothing, even though their reasons differ. Generation Swing does not buy second-hand primarily for hygienic reasons:

> I often go to second-hand stores, but I can never think of buying clothes there. I do not want to put on something that someone else has used before me.
>
> *(Rut, 73)*

Experiences from primary socialization may explain this since Generation Swing grew up during a period when hygiene and cleanliness were extensively discussed. They associate second-hand with the rag trade back in the old days.

Generation Y does not buy second-hand because they have a hard time finding something they like. Shopping for second-hand clothing usually involves a search process that takes time and it does not guarantee that the 'right' garments are found. From childhood, Generation Y has been able to choose among an abundance of cheap fashionable garments.

Environmentalists

As indicated above, both generations buy mainly second-hand clothing. In addition to environmental reasons, prices are lower as well as the possibility of finding higher quality at a lower price than if purchased new. Another advantage is that second-hand clothing relies less on trends and fashion:

> It's then often easier to find [second-hand] clothes that fit with what you have at home. Otherwise, it is easy to buy half a closet at one time and then come home and find that it does not fit with everything else. Since it's a new season and a new trend you have to throw away everything else.
>
> *(Elin, 28)*

Both generations sometimes find it difficult to find exactly what they are looking for when shopping for second-hand. Second-hand stores sometimes contain large amounts of clothing and can therefore be perceived as a bit messy. Also, clothing is not sorted in the same way as in regular stores. Furthermore, it can be difficult to plan in advance what to buy, because the assortment varies:

> You have to have more patience when buying second-hand. You can't decide from the start what you want. Instead you have to go and look for everything you could possibly need and then you can find what you need sometimes.
>
> *(Elin, 28)*

The attitude towards second-hand in society has changed over time and become more acceptable during recent decades. Generation Y has learnt during primary socialization that second-hand can be associated with unique clothing and environmental concern. Generation Swing has learnt this primarily during secondary socialization.

Care of clothing

Non-environmentalists

Generation Swing keep their clothing longer. One reason could be that they buy less fashionable clothing and instead focus on functionality. Another reason could be that they mend worn clothing by having access to appropriate equipment (e.g. a sewing machine) and knowledge of how to do it:

> I mend knees on my grandchildren's jeans. I have also mended stockings. A while ago, my husband bought alpaca socks. After a while the heel had a hole and I patched them with a patch that I had knitted.
>
> *(Margareta, 73)*

> I never mend anything, much because I lack the knowledge. I don't know how and I don't own a sewing machine. If it's a real crisis I'll probably call mum. Usually I often buy new [clothes].
>
> *(Linda, 31)*

Generation Swing's propensity to use clothing longer and to mend it can be traced back to primary socialization. They learnt this at home and at school when growing up. Apart from economic scarcity, they were taught to take care of their clothes, buy clothes of higher quality that lasted longer and mend clothes when needed. Even their experiences of rationing during the Second World War may have had an effect on their consumption of clothing. It is unthinkable for them to throw fully usable clothing in the dustbin. Generation Y has not experienced the same emphasis regarding the importance of learning how to care for clothing, neither at school nor at home. Learning consumer skills overall has been somewhat downplayed at school during recent decades, but can of course vary at home.

Environmentalists

Both generations use some of their garments for a very long time. To make their clothing last longer, they do not wash them as often, but let the garment rest between uses, air clothing and, if necessary, re-hydrate them to restore the fit. They first inspect each garment to assess whether the clothing can be used more times before it is washed. Another way they reduce the wear on clothing is to not use the dryer but instead hang clothes to dry. This also saves energy. Some of them hang-dry clothes even when it is below freezing outside, as long as the humidity is low:

> I live in the countryside, so we air dry all things. It works fantastically well, the wind just blows. During this dry period, it works even if it's cold, because there is no humidity. So if you hang it out now it dries very fast. So we use no energy to dry clothes.
>
> *(Jan, 68)*

Both generations mend their clothing, either by themselves or by asking someone else to do it. If mending is not possible, they turn the garment into something else, in other words redesign it. Both generations own the necessary equipment (e.g. sewing machine) for mending clothing, and possess good knowledge of how to mend. Several from Generation Swing described how clothing was repaired or made into something else when they grew up:

> It has of course changed a lot now. In my childhood, they turned the shirt collars and made changes on costumes and stuff.
>
> *(Rolf, 69)*

> When I was really little my mom took my grandfather's old jacket to the tailor to recut it to a small child's coat.
>
> *(Kurt, 71)*

One consumer in Generation Y explicitly stated that she appreciates that some clothing companies have started to mend clothing for customers:

> I have mended jeans. That's why I find it so amazing that Nudie [jeans company] now have begun to mend jeans. I've let them do it once and it turned out great. As for other clothes I sometimes sew on buttons and such stuff. I avoid trying to mend holes in for example wool sweaters.
>
> *(Anne, 31)*

As discussed above, Generation Swing has, regardless of being environmentalists or not, learnt to take care of clothing during primary socialization. This was not prioritized when Generation Y was growing up and, therefore, the environmentally conscious Generation Y has learnt this during secondary socialization.

Disposal of clothing

Non-environmentalists

Both generations throw away ragged and torn socks and underwear. However, some consumers in Generation Y also throw away other clothing. The reasons were that it is simple and convenient, but also that they did not want to give away clothing that could be perceived as imperfect to charity for reuse. Apparently, they were not aware that imperfect clothing can be recycled:

> I usually take the easiest and fastest way. How can I get rid of it quickly? If I have made up my mind, I would like to get rid of it. Often I'll go to those who collect clothes or throw it away.
>
> *(Linda, 31)*

Environmentalists

For most of the environmentalists, both Swing and Y, the main thing is that the clothing is not thrown away, but reused or recycled. Generation Swing explains their preference to reuse old garments by how things were done when they grew up:

> We grew up during a time when there was not that much, so we are probably affected by it. Also, I can't throw away food; it's really hard. It's the same thing with clothes that are still quite useful; it's not possible to throw away.
>
> *(Sonja, 77)*

Giving to charity

Non-environmentalists

Both Generations Y and Swing want to make sure that the clothing given to charity will actually benefit those in need:

> I can go to both Myrorna [charitable organization] and the Red Cross to look for clothes, but I never donate anything there. I don't want it to be sold. What I donate must go directly to those in need.
>
> *(Lisbet, 76)*

> To donate to charity is probably very deeply rooted. You feel better as a person. I'm very careful to check that it's really in good condition before I donate it. If there is some defect I prefer to throw it away.
>
> *(Linda, 31)*

It is clear that the philanthropic motive outweighs the environmental motive when choosing to give to charitable organizations.

Environmentalists

When the environmentally engaged generation Y and Swing decide to get rid of clothing, it is their environmental concern that determines to whom and where they give their clothing. They often place their clothing (both usable and worn out) in the nearest container for used clothing. The main thing is that the clothing is reused or recycled.

> Regardless of where the clothes are used and how they are used it is in all cases a resource that is taken care of. Each time a garment can be reused there are fewer resources used for new production.
>
> *(Jan, 69)*

For both generations, the respect for the environment outweighs the philanthropic motive when choosing recipients of the clothing they dispose of.

Trading and swapping clothing

Non-environmentalists

Some informants representing Generation Swing prefer to sell clothing of better quality to a private second-hand shop. Some informants representing Generation Y stated that they prefer to give clothing that they no longer desire to someone they know. One reason could be that clothing is perceived as very personal (Woodward 2005) and therefore given to friends.

Environmentalists

The two generations prefer to switch clothing with people they know. However, they do not hesitate to also exchange clothing with strangers. Most of them regularly join different clothing swapping events:

> Now it is getting so common this concept of swapping clothing. Only the last six months I have been involved in various friendship groups that have organized their own clothing swapping days.
>
> *(Linnea, 28)*

Regular exchange of clothes with strangers, for example visiting clothing swapping events, is something that separates the environmentalists from the non-environmentalists.

Discussion and conclusion

This study shows that consumer socialization during early ages is a critical construct for understanding how different generations experience and approach consumption of clothing. Generation Swing (non-environmentalists) have, during their adult life, maintained societal norms and values that they learnt during their upbringing. They show a more restricted orientation towards consumption of clothing, in terms of purchases, care and disposal. The meaning of consumption of clothing for them is more about having functional use, even though it does not exclude preference for aesthetics. The societal values emphasizing functionalism during the 1930s (e.g. Robach 2000) and the 1940s appear to have an effect on the Swing generation even today. This period was also characterized by a scarcity of economic resources (Mattsson 2010) forcing people to be thrifty. Therefore, Generation Swing (non-environmentalists) tends to prefer clothing that can be used longer. Consequently, they look for clothing of high quality with a classic design. In addition, they take good care of their clothing and, if necessary, mend it. The Swing (non-environmentalists) associate second-hand with the old-time rag trade, and therefore, they are less keen on buying second-hand.

The Generation Y (non-environmentalists) display more hedonistic consumer behaviour marked by the period when they grew up. It was a time characterized by an abundance of cheap fashionable clothing (e.g. Lindgren *et al.* 2005). Consumption for them is more about identity seeking. It is no longer a work that determines people's identity, but what they consume (Bauman 1998). According to the interviews, Generation Y (non-environmentalists) choose clothing primarily by criteria such as price, fashion and style. Generation Y (non-environmentalists) are, like Swing (non-environmentalists), reluctant to buy second-hand clothing, but for another reason, namely that it usually involves a long search process without any guarantee that the 'right' garment will be found. In general, it is easier to find cheap clothing in, for example, a fast fashion store. When they visit a second-hand store, they primarily do it to find unique garments. Generation Y (non-environmentalists) in contrast to Generation Swing (non-environmentalists), lacks the knowledge of how to mend clothing, partly because the incentives are low. The abundance of low-priced fashion during their upbringing has made care for clothing less important. In addition, they have not experienced the same emphasis in school regarding the importance of learning how to care for clothing. In grammar school, the subject 'home economics' encompasses 118 hours during nine years, and it does not exist in high school (Ekström and Larsson 2010).

Both Generations Swing and Y (non-environmentalists) indicate that philanthropic motives are important when giving clothing to charity. It is important that what is given will be used by people who need the clothing. Therefore, they prefer to give clothing to smaller charitable organizations that explicitly state what they do with the clothing they receive. Both generations (non-environmentalists) express scepticism towards organizations that sell clothing because they assume that their administrative costs will take a large share of the assets. This may be explained by the fact that most charitable organizations so far have highlighted the philanthropic motive when collecting clothing (e.g. Ekström et al. 2012; Hjelmgren and Gustafsson 2013) and environmental motives have not been emphasized in their marketing. Another similarity between Generations Swing and Y (non-environmentalists) is that they both throw ragged and worn out socks and underwear in to the dustbin. Some consumers among Generation Y (non-environmentalists) also throw away other clothing. One reason is that it is convenient, but it could also be that they do not want to give away clothing that can be perceived as imperfect, for example ragged and worn-out clothing.

Among the environmentalists, the differences between the generations are not as big as between the non-environmentalists. Both Generations Swing and Y (environmentalists) are trying to limit their consumption of clothing. When purchasing clothing, they often buy second-hand and if purchasing new clothes, high quality and durability are emphasized. In addition, they own the necessary equipment (e.g. sewing machine) for mending clothing and possess a good knowledge of how to mend. They also know other ways to extend the lifetime of their clothing (e.g. airing and hydrating) and thereby reducing the need to wash them. When they finally get rid of clothing, it is their environmental concern that determines to whom and where they donate their clothing. As a result, they often place their clothing (both usable and worn out) in the nearest container for used clothing. The main thing is that the clothing does not end up in the dustbin, but is reused or recycled. The similarity in behaviour between the generations (environmentalists) indicates that secondary socialization plays a role for both generations. The environmentally involved Swing generation have learnt about the benefit of acting environmentally consciously during their adulthood through secondary socialization. This new knowledge has changed their behaviour, for example, in the sense that they buy more second-hand clothes than Swing non-environmentalists. In a similar way, the Y generation (environmentalists) have learnt, during secondary socialization, the importance of quality and durability when choosing clothing, as well as how to take care of garments to increase their lifespan.

The study has shown that through secondary socialization it is possible to influence people to adopt a more environmentally conscious consumption pattern regarding clothing. The results also indicate that different issues need to be emphasized for the two generations depending on their various experiences. For example, while generation Swing needs to learn more about the potential of consuming second-hand clothing, Generation Y needs to learn more about functional needs, quality, durability and care of clothing. Both generations need to better understand the relationship between consumption of clothing and the effect on the environment, in particular the fact that giving clothing to charity is not only about helping people in need, but also about saving the environment (e.g. Ekström et al. 2012). If awareness increases that clothing is not only reused, but also recycled, the collection of ragged clothing is likely to increase.

In a consumer society facing limited natural resources, it is also important to consider primary socialization in future research. To what extent are young consumers of today learning about the effect consumption of clothes has on the environment? Also, the fact that a person may first become socialized and then become a socialization agent (e.g. Ekström 1995; Moschis 1976) is interesting to recognize in future research on environmental consumer socialization. Students may learn environmentally friendly behaviour in school that they later communicate to their parents and siblings. In a continuously changing society, we need to find out how and when different generations learn from each other about environmentally friendly consumption, including how to deal with waste. Also, different actors in society, such as politicians, authorities, retailers and schools, need to consider how learning about environmentally friendly consumption, including consumption of clothes, can be encouraged for consumers of all ages.

Acknowledgements

The authors would like to thank the Swedish Retail and Wholesale Development Council and the Swedish Research Council for Environment, Agricultural Sciences and Spatial Planning for funding this research project.

References

Allwood, Julian M., Sören E. Laursen, Cecilia M. de Rodríguez and Nancy M. P. Bocken (2006), 'Well dressed? The present and future sustainability of clothing and textiles in the United Kingdom', University of Cambridge, Institute for Manufacturing, Mill Lane, Cambridge CB2 1RX, UK. Available online at http://www.ifm.eng.cam.ac.uk/uploads/Resources/Other_Reports/UK_textiles.pdf (accessed August 6, 2013).

Bauman, Zygmunt (1998), *Work, Consumerism and the New Poor*, Buckingham: Open University Press.

Berger, Peter L. and Thomas Luckmann (1967), *The Social Construction of Reality*, Harmondsworth, UK: Penguin Books Ltd.

Brim, Orville Gilbert (1966), "Socialization through the life cycle," in Orville Gilbert Brim and Stanton Wheeler (eds), *Socialization after Childhood: Two essays*, New York: Wiley, 1–50.

Byun, Sang-Eun and Brenda Sternqvist (2008), 'The antecedents of in-store hoarding: measurement and application in the fast fashion retail environment', *International Review of Retail Distribution and Consumer Research*, 18, 133–47.

Carlson, Less and Sanford Grossbart (1988), 'Parental style and consumer socialization of children', *Journal of Consumer Research*, 15 (June), 77–94.

Carlsson, Annika, Kristian Hemström, Per Edborg, Åsa Stenmarck and Louise Sörme (2011), 'Kartläggning av mängder och flöden av textilavfall. SMED på uppdrag av Naturvårdsverket', [Mapping the amount and the flow of textile waste. SMED on behalf of the Swedish Environmental Protection Agency], Report No. 46. Norrköping, Sweden: Sveriges Meteorologiska och Hydrologiska Institut [Swedish Meteorological and Hydrological Institute].

Claudio, Luz (2007), 'Waste couture', *Environmental Health Perspectives*, 115(9), 440–54.

Churchill, Gilbert A., Jr. and George P. Moschis (1979), 'Television and interpersonal influences on adolescent consumer learning', *Journal of Consumer Research*, 6 (June), 23–5.

Ekström, Karin M. (1995), *Children's Influence in Family Decision Making: A study of yielding, consumer learning and consumer socialization*, Göteborg: BAS förlag.

— (2006), 'Consumer socialization revisited'', in Russel W. Belk (ed.), *Research in Consumer Behavior*, Vol. 10, Oxford: Elsevier Science, 71–98.

— (2010), 'Consumer socialization in families', in David Marshall (ed.), *Understanding Children as Consumers*, London: Sage, 41–60.

— (2013), 'Om behovet av konsumtionskritik i ett konsumtionssamhälle' [The need for critique of consumption in a consumer society], in Lennart Weibull, Henrik Oscarsson and Annika Bergström (eds), *Vägskäl, 43 kapitel om politik, medier och samhälle, SOM-undersökningen 2012 [Crossroads, 43 Chapters About Politics, Media and Society, The SOM-survey 2012]*, SOM-Institute, University of Gothenburg: SOM-report 59, 369–85.

Ekström, Karin M. and Gunnar Larsson (2010), 'Unga shoppar sig till status – arbetet har tappat sitt värde, debattartikel', *Dagens Industri*, October 27.

Ekström, Karin. M. and Nicklas Salomonson (2012), *Nätverk, trådar och spindlar: Samverkan för ökad återanvändning och återvinning av kläder och textil [Networks, Threads and Spiders: Cooperation for increased reuse and recycling of clothing and textiles]*, Vetenskap för profession [Science for the Professions], No. 22, Borås: University of Borås.

Ekström, Karin. M. and Nicklas Salomonson (2014), 'Reuse and recycling of clothing and textiles – a network approach', *Journal of Macromarketing*, 34(3), 383–399.

Ekström, Karin M., Eva Gustafsson, Daniel Hjelmgren, and Nicklas Salomonson (2012), *Mot en mer hållbar konsumtion: En studie om konsumenters anskaffning och avyttring av kläder [Towards a More Sustainable Consumption: A study of consumers' acquisition and disposal of clothing]*, Vetenskap för profession [Science for the Professions], No. 20, Borås: University of Borås.

Fletcher, Kate (2008), *Sustainable Fashion and Textiles, Design Journeys*, London: Earthscan.

FTI (Förpacknings- och tidningsinsamlingen) (2013), Statistic, http://www.ftiab.se/180.html (accessed 14 August 2014).

Grönhöj, Alice (2007), 'Green girls and bored boys? Adolescents' environmental consumer socialization', in Karin M. Ekström and Birgitte Tufte (eds), *Children, Media and Consumption: On the front edge*, Göteborg University: Nordicom, The International Clearinghouse on Children, Youth and Media, 319–33.

Hjelmgren, Daniel and Eva Gustafsson (2013), *Textilreturen i Ullared – ett experiment om återvinning [The Textile Return in Ullared – An experiment about recycling]*, Vetenskap för profession, [Science for the Professions], No. 25, Borås: University of Borås.

Lindgren, Mats, Bernhard Lüthi and Thomas Fürth (2005), *Me-We-generation: What business and politics must know about the next generation*, Stockholm, Sweden: Bookhouse Publishing.

Mattsson, Helena (2010), 'Designing the reasonable consumer: standardisation and personalisation in Swedish functionalism', in Helena Mattson and Sven-Olov Wallenstein (eds), *Swedish Modernism, Architecture, Consumption and the Welfare State*, London: Black Dog Publishing, 74–99.
McDaniel, Roxanne and Carole Bach (1996), 'Focus group research: the question of scientific rigor', *Rehabilitation Nursing Research*, 5(2), 53–9.
Moschis, George P. (1985), 'The role of family communication in consumer socialization of children and adolescents', *Journal of Consumer Research*, 11 (March), 898–913.
— (1976), 'Acquisition of the consumer role by adolescents', Unpublished doctoral dissertation, Madison: Graduate School of Business, University of Wisconsin.
Morley, Nicholas, Stephen Slater, Stephen Russell, Matthew Tipper and Garth D. Ward (2006), 'Recycling of low grade clothing waste', prepared by Oakdene Hollins Ltd, the Salvation Army Trading Company Ltd and Nonwovens Innovation & Research Institute Ltd., Defra Contract Reference: WRT152.
Morley, Nicholas, Caroline Bartlett, and I. McGill (2009), "Maximising reuse and recycling of UK clothing and textiles: A report to the Department for Environment, Food and Rural Affairs," Oakdene Hollins Ltd. Available online at http://randd.defra.gov.uk/Document.aspx?Document=EV0421_8745_FRP.pdf (accessed March 4, 2014).
Nordström, Ludvig. (1938), *Lort-Sverige*, Stockholm: Kooperativa förbundets bokförlag.
Parment, Anders (2008), *Generation Y*, Malmö: Liber.
Robach, Cecilia (2000), *Ting för den moderna människan - Utopi och verklighet, svensk modernism 1900–1960* [Things for the modern man – utopia and reality, Swedish modernism 1900–1960], Stockholm: Norstedts förlag.
Roos, John M. (2010), Konsumtionsrapporten 2010. [The Consumer Report 2010]. School of Business, Economics and Law, University of Gothenburg: Center for Consumer Science.
Sheahan, Peter (2005), *Generation Y: Surviving (and thriving) with Generation Y at work*, Praharan, Australia: Hardie Grant Publishing.
Sullivan, Pauline and Jeanne Heitmeyer (2008), 'Looking at Gen Y shopping preferences and intentions, exploring the role of experience and apparel involvement', *International Journal of Consumer Studies*, 32(3), 285–95.
Tallman, Irving, Ramona Marotz-Baden and Pablo Pindas (1983), *Adolescent Socialization in Cross-Cultural Perspective: Planning for social change*, New York: Academic Press.
Ungerth, Louise (2011), *Vad händer sen med våra kläder?* [*What Happens Then with Our Clothes?*], Stockholm: Stockholm Consumer Cooperative Society.
Ward, Scott (1974), 'Consumer socialization', *Journal of Consumer Research*, 1, 1–16.
Woodward, Sophie (2005), 'Looking good: Feeling right – aesthetic of the self', in Susanne Küchler and Daniel Miller (eds), *Clothing as Material Culture*, Oxford: Berg, 21–39.

10

UNPACKING CORPORATE SUSTAINABILITY

Sustainable communication, waste and the 3Rs in a network society

Pierre McDonagh and Andrea Prothero

Introduction

Examining the 3Rs asks us to examine the disposition of waste and, as de Coverly *et al.* (2008) discuss, social engagement is the key ingredient in tackling, what they refer to as the *Hidden Mountain* of waste:

> to theorize waste as simply a waste management issue or indeed as a marketing or a consumer research issue is problematic in itself. Rather it is a societal problem and one that is culturally embedded.
>
> *(de Coverly* et al. *2008: 300)*

As the authors remind us, disposition has not enjoyed as much attention in the consumption literature as the acts of buying 'stuff' and engaging in consumption practices have. While the authors stress that as a society we are uncomfortable discussing waste, they also emphasise that waste has not suddenly become an important issue, as Vance Packard's *The Waste Makers* published in 1960 clearly illustrates. In recent years we have witnessed further engagement with the waste issue within consumer research (Fuller *et al.* 1996; Peterson 2006), and have seen papers on topics such as sharing (Belk 2010; Bardhi and Eckhardt 2012; Ozanne and Ozanne 2011), consumer waste initiatives (Dobscha *et al.* 2012), and the reuse of our waste (Brosius *et al.* 2013), and the issue of collaborative consumption has taken centre stage in wider consumer discourses (Botsman and Rogers 2010).

At the same time, we have had discussions focusing on how sustainability strategies are communicated with key stakeholders, and in the early 1990s there was much discussion around corporate greenwashing (Landler 1991; Kangun *et al.* 1991), which subsequently became a term used by NGOs and pressure groups who were keen to point out the poor sustainability activities of organisations, particularly global corporations. Of late, organisations have embraced sustainability in what many suggest is a more meaningful way than in the early 1990s. But can we realistically talk of sustainable communication being at play? Is consumer and corporate engagement with the 3Rs indicative of a radical sea change in society? The swing to the digital within our network society (Castells 2010) has opened up a pathway to an even greater cacophony around all things 'sustainable'. The work of Manuel Castells spans the 1970s to the present and the author argues that in all sectors of society we are witnessing a transformation in how constitutive processes are organised. It implies we are moving away from 'hierarchies' to 'networks'. Such a transformation is as much an organisational as a cultural question. This, Castells argues, started in the 1970s through the confluence of three independent trends – the invention of microelectronics and the IT revolution; the crisis of industrialism in both capitalist and *statist* societies; and the profound cultural challenge mounted by the rise of counter-cultural social movements in the late 1960s. Within the network society, information and scrutiny can become quasi instantaneous. People 'google' facts, fun and firms from their smart phones in a matter of seconds. As a consequence, we are more technologically capable of asking questions of firms, which has accidentally led to a partial operationalisation of the theory of sustainable communication.

McDonagh (1998: 599) defined this process as 'an interactive social process of unravelling and eradicating ecological alienation that may occur between an organisation and its publics or stakeholders' – Figure 10.1 outlines the process in a more conventional understanding of communications.

One uneasy question that needs to be asked is whether sustainable communication proper is at play or has a prevailing corporate order merely appropriated the semblance of the sustainable communication process for its own agenda? It is noted elsewhere in the works of Giddens (1986) and Milbrath (1984) that the dominant system has an ability to absorb oppositional forces to its own end (see also Kilbourne et al., 1997). This was previously discussed in the journal *Consumption Markets & Culture* using the prism of the Hegelian Master/Slave dialectic (Desmond et al. 2000). Consequently, while it is hugely problematic to answer whether sustainable communication has been appropriated, this still forms an aspect for this chapter by considering the relations between corporations, consumers and citizens as they collectively engage with the 3Rs. Many corporations are now describing themselves as sustainability solution providers and are informing consumers of the measures they are taking to help with environmental challenges; and they are using their company websites and social media to engage in sustainability conversations and information provision with their consumers. The stories being communicated suggest that waste is now important and that consumers can participate in helping the environment though the 3Rs, indeed, as we demonstrate below, there is recognition of waste as a societal problem within these communications.

This chapter explores some examples of what is currently happening within the fashion and apparel market in relation to the 3Rs, before further considering

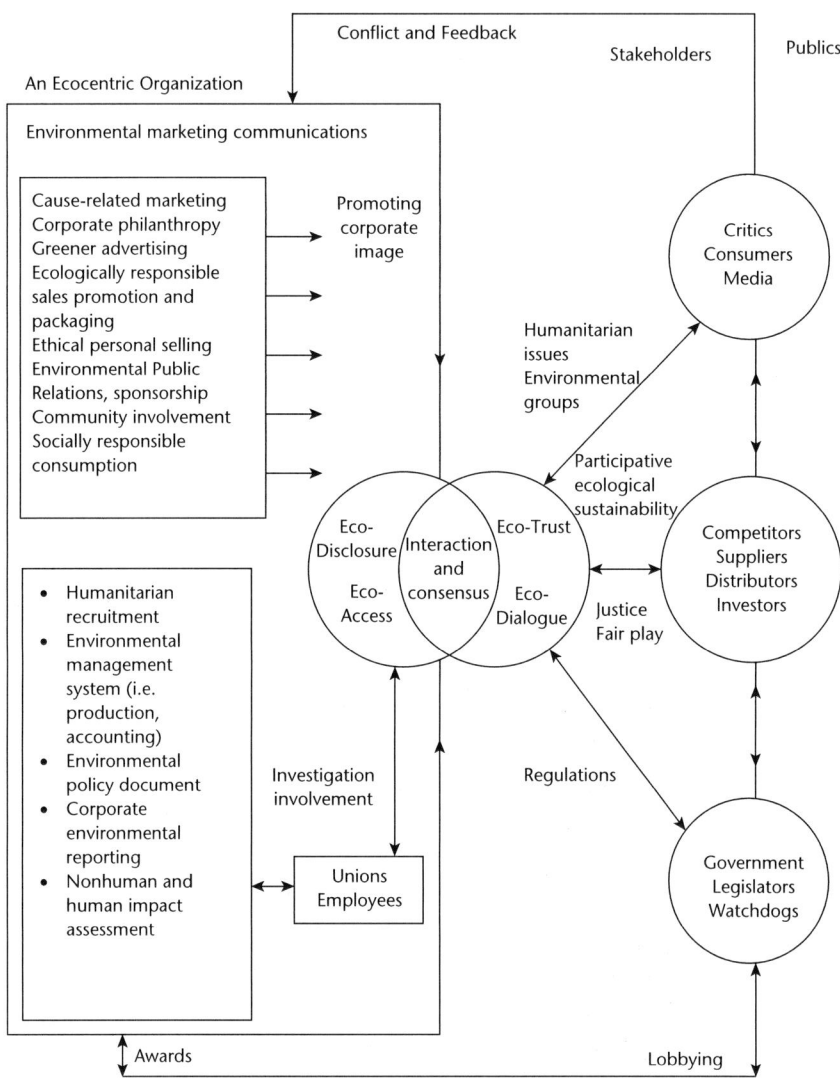

FIGURE 10.1 The process of sustainable communication. Source: McDonagh (1998)

how these strategies can be related to the broader issues surrounding the systemic disposition of waste. Within the fashion industry are we beginning to see new conversations about waste opening up, and are we beginning to understand waste differently? Are we witnessing changes to the 'systemic logic that seeks to minimize public displays of waste' (de Coverly *et al.* 2008: 299)? Have corporations actively developed significant waste policies and are they engaging in sustainable communication to help facilitate these changes?

Fashion and the 3Rs – examples from website communications

The fashion industry has been criticised on many different levels in recent decades – be it in relation to the use of sweat shops to produce clothes, the use of children as labour in clothing factories, or the exploitation of workers through poor pay and conditions – as well as being criticised because of the throw away and cyclical nature of the industry itself. The industry has responded with both individual company responses and collective initiatives, such as the Nordic Initiative Clean & Ethical (2013) and the Sustainable Apparel Coalition (SAC; 2013). The SAC has over a hundred members, drawn from industry members such as Adidas, Nike, Levi's, H&M and Walmart, and also includes NGOs, government and education members, such as Fairtrade International and the World Resources Institute. Sustainable fashion, as a concept, is currently being discussed within the fashion field (Fletcher 2008; 2010), as well as taking traction in disciplines such as the supply chain (de Brito *et al.* 2008) and retailing (Choi and Chiu 2012). We explore below examples from the fashion and apparel industry of three companies' waste policies in the 3Rs arena and how they communicate these (via company websites and social media) with their consumers – Nike, H&M and Marks and Spencer. The landing pages of each of the organisations were visited in October/November 2013. Due to copyright restrictions, the authors are not able to publish these landing pages here but urge the reader to view the sophisticated and seductive representations of the 3Rs found there. What follows is a discussion of each of the landing pages and URLs to current pages (June 2014) are included in the references.[1]

Nike

The initial landing page for Nike sees no mention of sustainability. However when you google 'Nike sustainability' you are directed to a separate landing page (http://nikeinc.com/pages/responsibility). The page has a number of different links which include: Sustainable Business at Nike Inc with sub links to Designing Products, Manufacturing, Our Impacts, Reporting and Governance; Nike Sustainable Business Performance Summary; and 'A shoe. A company. A journey'. When you click on the Nike Sustainable Business Performance summary you are provided with more details about the 'shoe, the company, the journey' and its strategy regarding closing the loop. A short music video about the company's achievements with its 'Nike Grind' and the reconstituted materials for the Reuse-A-Shoe programme, is available for viewing.

Under the 'targets and commitments' link, the company highlights reducing waste, cutting energy, rejecting toxics and 'raising the bar', and refers the viewer to its detailed Corporate Responsibility report. In this report, there are impressive claims and related statistics within the text:

> Across our whole value chain, our single greatest waste item is corrugated cardboard. We've worked for years on reducing the overall content in our boxes

and developed alternatives with various degrees of success for recyclability and performance. Our boxes are sustainable with 100% recycled content since 1995 and 15% waste removed from 1995 to 2006, but with the advances we've made to date it's more and more difficult to make meaningful progress. But we're back at it – aiming for a 10% reduction in weight of our footwear shoebox.

We're also targeting reducing waste even further in footwear manufacturing, where we have already achieved a 35% reduction in grams of waste per pair in contracted footwear factories for the NIKE Brand over 10 years.

Distribution centers handle a high volume of waste but they do not generate the vast majority of it – most comes in from shipping and goes out based on customer requirements. Other areas – including the offices and retail environments – also are important to the overall waste picture. We focus our efforts in recycling, reusing, repurposing or composting as we also work to reduce the amount of waste.

(See more at http://www.nikeresponsibility.com/report/content/chapter/waste).

Elsewhere, viewers of the website are informed that the corporation takes plastic bottles and recycles them into sports clothing, and uses powerful imagery to confirm this. There is also an interesting video on the company's 'Reuse-A-Shoe' campaign.

The company website informs the consumer that it takes sneakers and uses them for reuse in materials for AstroTurf pitches,

> Worn out. Play on. Turn your old sneakers into places to play.
>
> Nike Reuse-a-Shoe takes worn out athletic shoes and grinds them down to create a new material called Nike Grind, which is used to make high-quality sports surfaces including courts, turf fields, tracks and more since 1990, we've transformed 28 million pairs of shoes and 36,000 tons of scrap material into Nike Grind for use in more than 450,000 locations around the world.
> *(http://www.nike.com/us/en_us/c/better-world/reuse-a-shoe)*

When we explore more, we are told that the company wants to close the waste loop,

> Nike is constantly striving for the best, creating value for business and innovating for a better world. Nike's vision is that our products will be "closed loop" – that is, they will use the fewest possible materials and be assembled in ways that allow them to be readily recycled into new products. Our long term vision is to create a continuous loop without waste. As a company, we are innovating materials and manufacturing processes as we work towards finding a solution to this complex issue. Examples of how this comes to life are our Reuse-A-Shoe and Nike Grind programs. Whether it's a worn-out shoe that's sliced and refined or repurposed manufacturing scrap, the end results are the same. Sustainable, long-lasting materials designed for professional level performance.
> *(http://www.nike.com/us/en_us/c/better-world/reuse-a-shoe)*

As well as the graphics on Nike Grind, the company also has its 2010/11 eighty-five page Sustainable Business Report available for download which has detailed explanations and graphics to explain how it is performing across its various impact areas, and in particular in relation to the 3Rs. It would appear that Nike's sustainability strategies and their communications policies show us how their actions benefit society in terms of buildings, game play and parks and recreational activities and allows Nike to continue to contribute to the problem within a consensus culture (de Coverly *et al*. 2008).

H&M

For H&M we witness a number of reduce, reuse, recycle initiatives as part of their sustainability reporting, which viewers to its website are invited to consider: The initial company landing page has a prominent link on 'sustainability', which has an extensive drop-down menu (http://about.hm.com/en/About/sustainability/commitments/reduce-waste.html).

There is an opening message from the CEO on the company's sustainability programme, as well as specific links to company videos and case studies in the sustainability field. The company also provides a list of seven sustainability commitments, which fall under the umbrella heading of H&M Conscious. The overall aim of H&M Conscious is included in the overall vision statement of the company that states, 'Our vision is that all business operations shall be run in a way that is economically, socially and environmentally sustainable.' The company presents its sustainability report for 2012 as *Conscious Actions* and has striven to close the waste loop on textile fibre use. Following a pilot scheme in Switzerland, it claims to be the first company in the world to offer its customers the chance to bring any 'unwanted clothes from any brands' to any of its stores in the 53 markets where it operates for recycling. It reuses or recycles these, nothing goes to landfill. The seven commitments highlighted focus on the following key areas, where each is discussed at length on the website:

1. Provide Fashion for Conscious Consumers
2. Choose and Reward Responsible Partners
3. Be Ethical
4. Be Climate Smart
5. Reduce, Reuse, Recycle
6. Use Natural Resources Responsibly
7. Strengthen Communities

(http://about.hm.com/en/About/sustainability/commitments/our-seven-commitments.html)

Furthermore, it encourages all customers to use recycled plastic bags and uses recycled materials, such as recycled wool and polyester, to make its clothes. For example, recycled polyester is often made of PET plastic bottles. In 2012, the equivalent of about 7.9 million bottles were used by the organisation. It handles 95 per cent of its waste in its distribution centre, with the ultimate aim of zero per cent waste. The key areas of 'Don't let fashion go to waste', Carrier Bags, Waste Management and Packaging each have their own drop-down menu and detailed discussions of the companies activities in each area.

Marks and Spencer

For Marks and Spencer, the initial web landing page has no mention of 'sustainability' but at the bottom of the landing page there is a link to Plan A (http://corporate.marksandspencer.com/plan-a).

The top of the web landing page has an emphasis on sales and the product range. The bottom of the initial landing page, has a link to 'Plan A' under the 'About Marks & Spencer' column. Marks and Spencer have been lauded for their Plan A (the company motto being, *because there is no Plan B)* initiative, introduced in 2007. Under the commitments of this campaign, the company is working with their customers and suppliers to tackle a number of sustainability challenges including climate change, waste, sustainable raw materials, ethical trade and healthier lifestyles for consumers. In terms of waste, since 2007, M&S reports it has cut back on its non-glass packaging across food, clothing and homeware by an average of 26 per cent. Clothing packaging has decreased by 46 per cent, including reductions achieved through hanger recycling. In 2012, M&S achieved their target of sending no waste to landfill from their stores, offices and warehouses in the UK and Republic of Ireland. The company ensures different types of material go to the most appropriate type of reuse or recycling.

Under the Shwopping initiative, promoted using the British actress, Joanna Lumley, the company has partnered with the charity Oxfam to enable customers across the nation to resell, reuse or recycle their clothes. The scheme aims to reduce the volume of clothes that are thrown away to landfill, reducing the environmental impact, while also supporting Oxfam's many good causes. Customers can bring unwanted pieces of clothing and put them in bins known as 'Shwop Drops'. The clothes are then given to Oxfam to reuse, recycle or resell. To date nearly four million clothes have been donated and over £2 million has been raised for charitable causes (Hamed 2013), with the initiative also contributing to the company's zero-waste policy. This is detailed under the 'Doing the right thing' section of the 'Plan A' page on the company website.

In October 2012, M&S sold its first ever 'shwopped' ladies coat, a black jacket created from recycled material made from shwopped clothes. The Shwop coat follows the 'world's most sustainable suit'. The Shwopping initiative earned M&S a 'Big Society Award' from the British Prime Minister in April 2013 (sustainablereview.net), M&S were congratulated for the 'simple and innovative idea' by David Cameron. Adam Elman, Head of Plan A Delivery at Marks and Spencer, said:

> We're delighted to be recognised by The Prime Minister for Shwopping. At M&S we're passionate about giving used clothes a future. One man's old shirt

is another's retro classic and we can even transform tatty clothing into new products, such as mattress or car seat fillings.

There's no excuse for sending textiles to landfill and that's why we're doing everything we can to make it easy for people to recycle with Oxfam. So far we've recycled over four million garments this year and have the ultimate aim of recycling one piece of clothing for every one we sell, over 350 million a year.

(http://sustainablereview.net/ ms-shwopping-campaign-wins-big-society-award/)

Tales from the other side – criticisms of Nike, H&M and Marks and Spencer

Since the early 2000s, amidst criticisms of company greenwash strategies, are we now witnessing a situation where waste has been recognised as a societal problem, and companies are engaging with consumers to address the cultural embeddedness of waste and provide solutions to the systemic waste problem? How have the internet and social media sites been used by NGOs, pressure groups, academics, citizens and governments to critique the company strategies so eloquently presented by companies in their communications with consumers?

Nike

The discussion above provides examples of the various 3R initiatives adopted by the three companies. On the face of the communications of company sustainability strategies, we could argue that something wonderful seems to have happened since Nike was accused in 2001 of violating the Global Compact principle. Tim Connor of NikeWatch in Australia reported that Nike had repeatedly failed to uphold the principle of free association and the effective recognition of the right to collective bargaining in China, Indonesia, Thailand, Cambodia, Mexico and elsewhere (O'Connor 2001).

A legal example illustrates just how difficult it is to assess whether or not a company is making misleading or deceptive claims. In *Nike v Kasky*, Mark Kasky a former basketball player and now a director of Green Century Institute filed a lawsuit against the corporation following newspaper adverts and several letters Nike distributed when its working conditions in its overseas factories were heavily criticised. Kasky claimed Nike was engaging in false advertising. Nike on the other hand claimed that it was entitled to make these claims and the advertising laws did not the cover their right of expression under the First Amendment. The local court agreed with Nike but then the Californian Supreme Court overruled and claimed that the corporation's communications were 'commercial speech' and therefore subject to false advertising. Although the US Supreme Court agreed to review the case, it sent the case back to trial without making any substantive ruling on the constitutional issues. However, before any findings were made on the accuracy of Nike's statements the parties settled out of court, which left the California

Supreme Court's denial of Nike's immunity claim as precedent. While Goldstein (2003: 79) concluded that:

> It is deeply unfortunate that the U.S. Supreme Court did not seize the opportunity presented by Nike v. Kasky to bring greater coherence to the commercial speech doctrine. But until the justices return to the issue, the lower courts should recognize that the California Supreme Court's decision in the case is unpersuasive and destined ultimately to be rejected.

Volokh (2003: A16) observed that:

> the problem remains unresolved as businesses, unfortunately, don't have much of an answer to the question 'when may businesses be sued for making allegedly false or misleading claims' – except to be extra careful about any statement that some might construe as misleading, especially in California.

He clarifies how the use of commercial speech is to be viewed,

> In most public debate, honest mistakes and statements that are true but misleading are constitutionally protected. A journalist or scientist who writes 'Studies show product X cures baldness' can't be sued, at least unless his statement is a knowing lie. If speakers could be sued based on allegedly misleading assertions, the risk and cost of litigation would often deter them from saying even true things. So under the First Amendment, the remedy for error is rebuttal, not litigation.

> But if the same statement appears in an ad for product X, the advertiser can be held legally liable. If you warrant something about your product, your warranty better not be mistaken or misleading. This is traditionally referred to as the 'commercial speech' exception to full First Amendment protection: Accurate 'commercial speech' is usually constitutionally protected (though not as much as other speech), but false or misleading commercial speech is unprotected.

> 'Commercial speech' has always been an ambiguous label. It sounds like speech sold in commerce, or speech about commerce, but that's not right. This newspaper qualifies on both those counts, and it's fully protected speech, not commercial speech. Rather, the commercial speech doctrine is generally limited to speech that directly or indirectly proposes a commercial transaction with the speaker – commercial advertising, more or less. A billboard saying 'Buy Nike shoes: Made with high-paid labor,' for instance, would be commercial speech. So, honest mistakes and misleading statements in public debate are protected by the First Amendment. Similar statements aimed at getting people to buy your products are unprotected.

Nike is one of those companies that has been constantly criticised for its broad corporate responsibility activities, as the two examples above demonstrate. In the

2000s, its alleged sweat shop policies were cleverly depicted in the Adbusters parody ad. The ad shows a picture of a barefoot girl running with the following text, accompanied by the Nike logo and swoosh symbol,

> You're running
>
> Because you want that raise,
>
> To be all you can be.
>
> But it's not easy
>
> When you
>
> Work
>
> Sixty hours a week
>
> Making sneakers in an
>
> Indonesian factory
>
> And your friends
>
> Disappear
>
> When they
>
> Ask for a raise.
>
> So think
>
> Globally before you decide
>
> It's so cool
>
> To wear
>
> Nike
>
> *(https://www.adbusters.org/content/nike-running)*

How then have its current 3Rs policies stood up to scrutiny? In 2011, Greenpeace launched its successful *Detox* campaign, using social media sites such as Twitter, to call on clothing and apparel companies to help eliminate water pollution due to the use of toxic chemicals in the fashion industry. H&M, Nike and Marks and Spencer all signed up to the campaign and agreed to eliminate hazardous chemicals from their supply chains and the full life-cycle of their products by 2020. In October 2013, Greenpeace applauded H&M for its continued actions, but criticised Nike for needing to do more work. The NGO emphasised that Adidas, Nike and Li-Ning were guilty of greenwashing and they had not followed through on their detox commitments, both Adidas and Nike responded to the criticisms and highlighted how they felt they were addressing the challenges to which they had committed (Giegerich 2013).

H&M

While H&M were applauded for their detox commitments, the company has been criticised for its broader corporate responsibility acts in Bangladesh (H&M is the largest buyer of clothes in Bangladesh). Concerns were raised over its slowness to react to the Bangladeshi factory collapse (Lo 2013). Even though the company itself was not associated with the factory, NGOs were quick to point out that the company should be doing more for Bangladeshi workers. In September 2013, along with Lidl and Gap, it was reported that workers for the company were engaged in 15-hour shifts (Butler 2013). A Google search for 'H&M Bangladesh' drew 4,970 results in November 2013, highlighting that corporate actions are indeed scrutinised and discussed at length online. On one sustainability website, H&M is criticised for its practices drawing parallels with its sustainability and conscious actions,

> 'H&M claims that [its] clothes are made with responsibility for people and environment, but hundreds of overworked and malnourished workers faint during their daily work,' says Christa Luginbühl, a coordinator for the Clean Clothes Campaign says. 'A fashion collection cannot be "conscious," "sustainable," or "responsible" if a producer denies garment workers the basic human right for a living wage.'
>
> *(TodayEco.com 2013)*

What the H&M example clearly demonstrates is that it is no longer possible for companies to make sustainability claims without these being considered alongside broader corporate responsibilities, and vice versa.

The Avaaz organisation takes H&M to account with its own use of marketing communications via a protest campaign. Using similar imagery and language to H&M's website, Avaaz speaks directly to Karl Johan, CEO of H&M asking him if there have been 'Enough Fashion Victims' (Lo 2013).

It is claimed that it was only following this protest campaign that H&M started working with the Bangladeshi government to raise the minimum wages of textiles workers, which have since gone up by 81 per cent. The Bangladesh Development Plan is now clearly visible on the company website.

Marks and Spencer

As noted above, Marks and Spencer launched its Plan A initiative in 2007, and introduced a second 2010–2015 plan. The company's 2013 report has been well received by environmentalists, and the general perception appears to be that the overall sustainability policies put in place by the company are welcomed as being concrete and actionable. At the launch of its 2013 Plan A report, Jonathon Porritt, of the Forum for the Future Board, a respected environmentalist, and also member of the M&S sustainability board, had mostly praise for the sustainability polices of

M&S, and only mentioned a very small number of areas where he felt M&S could do better (Hamed 2013). However, some of its activities have been criticised by other NGOs. The 2012 Shwop campaign for instance, received an average rating of 2.7 by the Greenwashing Index (2012) – a score of 1 means the campaign is 'authentic' and 5 suggests 'bogus'. While M&S won a UK government Big Society award for its Shwopping campaign in 2013, there are numerous criticisms of the campaign for not tackling bigger issues, such as slow fashion (Keithpp's Blog 2012). As one blogger comments:

> Shwopping is a slick marketing campaign to encourage easily led fools to empty their wardrobes and run off down to M&S to buy more clothes. Green it is not.

Others have also criticised the campaign:

> On the other hand, this initiative is about becoming less bad, and as Bill McDonough said, 'being less bad is not being good; it's being bad, just less so – by definition.' It means that shwopping, unlike swapping for example, doesn't take a holistic approach of the problem and avoid the questions concerning the excessive consumption of garments, which might be the main problem in the first place.
>
> *(Godeinlk 2012)*

Following the launch of its sustainable suit, as mentioned above, some journalists criticised M&S and other fashion houses. One journalist (Freeman 2012) asked, 'Is the Fashion Industry Blinding Us With Greenwashing'. The author lambasted M&S on Twitter, and stated that developing a sustainable suit did not mean anything when the company had been recently accused of human rights abuses by Anti-Slavery International and the Centre for Research on Multinational Corporations (SOMO) (Anti-slavery 2012), with the group commenting,

> 'Slavery on the high street' finds five Indian clothing manufacturing companies supplying leading European and US brands, using the forced labour of young women and girls made to live in prison-like conditions in cotton spinning mills and factories around Tirupur, western Tamil Nadu.
>
> *(Anti-slavery 2012)*

Other criticisms were visible via Twitter, with one tweeter (@davekinkydevils) commenting 'This is not just slave labour, this is M&S slave labour. #workfare.'

DEFRA (The Department of Environment, Food and Rural Affairs in the UK) recently published a report on recycling and reuse in the textile industry, and found that while swapping sites online are growing, and are helpful in raising awareness of fashion waste, the amount of clothes swapped in tonnage terms is very low (Morley *et al.* 2009). At the same time M&S has been criticised, along with other companies operating on an 'ethical platform', as having double standards as a result of its partnerships with major oil companies BP, Shell and Chevron (Miller 2012).

Discussion

Of the three companies explored, H&M initially gives the clearest directions on its sustainability and 3Rs strategies for viewers of its webpages. However, once Nike and M&S are explored further they also have detailed waste stories to share on their websites. While it could be argued that the sustainability strategies and subsequent communication of these by large companies seems to confirm a consensus culture between companies and their consumers, it may be too early to make such a claim. The critiques of companies might also suggest that they are operating under what Desmond (2003) labels 'liberal pluralism' and would deem to be exhibiting enlightened self-interest when it comes to sustainability. Most have adopted the common metaphor of being 'on a journey' which is a difficult thing to argue against. As Sturdy and Grey (2014) observe of organisational change management, it could be suggested that there is a mechanistic desire to manage the sustainability issue. Similarly, not all consumers desire or are interested in company sustainability policies so they may be happy to accept 'the commercial speech' and the organisation is happy for them to do so. In this sense, commercial speech forms another part of the marketing armoury, which also has the capacity to be augmented with diagrams of eco-efficiency, videos of recycling competencies and strong imagery. If one considered a Baudrillardian analogy the enquiry, or person seeking sustainability answers, runs the risk of becoming seduced by the spectacle. This might explain why M&S and Nike do not foreground sustainability on their landing pages, but have the relevant information for the more eco-curious person to find.

All three organisations seem to use the eco-efficiency argument in their online communications to consumers, as this seems to be a convenient place to tell good news stories about what the company has achieved. Trickier legal issues are not always discussed or acknowledged on the websites (although H&M is an exception, albeit a slow one, here with respect to minimum wages in Bangladesh). For some, this in itself may be enough to suggest that enlightened self-interest is at play. Bear in mind here that if there was greater disclosure then consumers might well adopt the 'refuse to consume' approach which is depicted as a higher level of sustainability engagement in The Inverted Pyramid of Sustainability (TIPS; Dobscha *et al.* 2012). TIPS tackles some of the more systemic issues of waste disposition, as well as exploring some of the easier 3R solutions.

Overall the assessment of the 3Rs strategies of the three companies within the fashion and apparel industry has shown how companies use some aspects of sustainable communication in their sustainability strategies with consumers, and how the network society has allowed them to do this, using both company websites and via social media interactions with consumers. At the same time, however, NGOs and sustainability pressure groups in particular, also use these same networks to highlight critiques of the broad corporate behaviour of organisations, including their sustainability

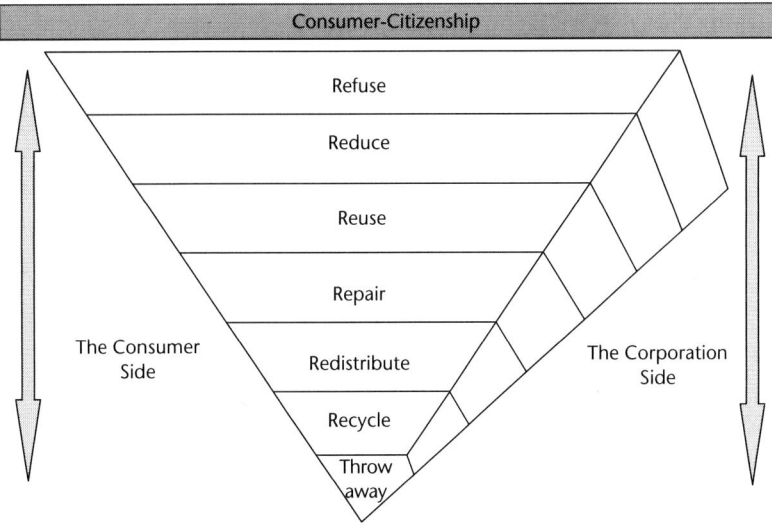

FIGURE 10.2 The inverted pyramid of sustainability

and 3R activities. We have presented some examples here of how the three organisations have been criticised for their 3Rs strategies and their broader corporate responsibilities. To this end, we have witnessed how 3Rs strategies, within a broader sustainability framework, do help organisations engage with the waste discourse, they do not, however, fully engage with the wider waste issues, as highlighted by Dobscha et al. (2012), nor engage in conversations with the 'elephant in the room' – namely the on-going 'overconsumption' of fashion items. It is these issues, which are scrutinised and discussed on the internet by NGOs, pressure groups and indeed the everyday citizen who may choose to rant about organisations via Facebook, Twitter or using various micro-blogging sites. In response to de Coverly et al. (2008) we have seen how the three organisations examined have considered how to dispose of waste at the systemic level, and how they have successfully communicated this with their consumers. This does not mean, however, that these companies are seen as 'sustainability champions' by everyone. Their 3R policies, along with wider sustainability and ethical concerns, are criticised because the systemic conversations have not been taken far enough. The sustainable communication process requires a much more radical engagement through the adoption of the four elements of eco-dialogue, eco-access, eco-disclosure and eco-trust than we are presently witnessing. It would seem that companies are still adopting what McDonagh (1998) describes as a procedural stance on the environment, which seems to fixate on the eco-efficiency story. Such an approach is not that surprising, as the industry needs to increase its competency to talk about sustainability in a meaningful manner. Not only does the process require this, but if the rise of the sustainability professional is 'managed' just to project a managerial bias then those people outside the firm will lose eco-trust. As a result, even some twenty years after its conceptualisation, the theory of sustainable communication still remains a pipedream (McDonagh, 1995) for marketers whereas it ought to be expedited to increase eco-dialogue and work against a pure obfuscation that is commercial speech.

Thus, there is a failure to tackle the broad question of producing too much fashion, even if what is produced is more sustainable, which means opportunities for a green commodity discourse have not reached their full potential (Prothero and Fitchett 2000; Prothero *et al.* 2010). At the same time, the waste strategies of these companies, and indeed their wider corporate responsibilities, are being constantly challenged, which means greenwashing is increasingly difficult to get away with, and the communication strategies which are employed need to be carefully scrutinised as 'commercial speech'.

Overall, we see illustrations of industry, in this case fashion, having begun to consider waste at a systemic level, companies use online communications strategies to highlight their activities; and the power of stakeholders keeps them in check. The overall sustainability engagement generally, and in the waste area specifically, has advanced considerably since the initial discussions of recycling in the latter half of the twentieth century. As such, we argue that we are indeed moving towards a more interactive social process of unravelling and eradicating ecological alienation that may occur between an organisation and its public or stakeholders. However, it perhaps suffers from the same symptoms as noted within organisational change management by Sturdy and Grey (2014) pointing to alternatives perspectives as grounded in core assumptions of pro-change bias, managerialism and universalism. Consequently, while it is not fully present today, we cannot dismiss the future potential arrival of sustainable communication, but there needs to be a much more rigorous attempt at going beyond the use of mirrors, websites and corporate good news stories towards an evidential third-party endorsement of the sustainable brand in society. This is captured by Kilbourne (2004) as follows,

> Sustainable communication as conceptualized by McDonagh (1998), would contribute greatly to the development of sustainable consumption, but it offers a direct challenge to the DSP with liberalism as its basis. Because one of the essential features of all paradigms is their ability to deflect radical change, the prospects for sustainable communication are very limited. We cannot expect that the dominant groups will embrace it and incorporate it into the standard marketing strategy.

Based on the review provided in this chapter it would seem that Kilbourne's (2004: 204) observation of the challenge for sustainable communication is still apposite ten years later.

Note

1 All URLs were correct at the time of going to press; however, we cannot guarantee that they are still correct or that these pages are still available.

References

Anti-slavery (2012), 'Top UK high street brands selling clothes made through slavery'. Available online at http://www.antislavery.org/english/press_and_news/news_and_press_releases_2009/010612_ slavery_on_the_hight_street.aspx (accessed 9 November 2013).

Belk, Russell W. (2010), 'Sharing', *Journal of Consumer Research*, 36(5), 715–734.

Bardhi, Fleura and Giana M. Eckhardt (2012) 'Access-based consumption: The case of car sharing', *Journal of Consumer Research*, 39(4), 881–898.

Botsman, Rachel and Roo Rogers (2010), *What's Mine Is Yours: The Rise of Collaborative Consumption*, New York, NY: Harper Business.

Brosius, Nina, Karen Fernandez and Hélène Cherrier (2013), 'Re-acquiring consumer waste: Treasure in our trash?' *Journal of Public Policy and Marketing*, 32(2), 286–301.

Butler, Sarah (2013), 'Workers for Lidl, H&M and Gap in Bangladesh work 15-hour shifts', *The Guardian*. Available online athttp://www.theguardian.com/fashion/2013/sep/23/workers-in-bangladesh-long-hours (accessed 9 November 2013).

Castells, Manuel (2010), *The Rise of the Network Society: Information Age: Economy, Society and Culture Vol. 1* (Information Age Series), Chichester: Wiley-Blackwell.

Choi Tsan-Ming and Chun-Hung Chiu, (2012), 'Mean-downside-risk and mean-variance newsvendor models: Implications for sustainable fashion retailing', *International Journal of Production Economics*, 135(2), 552–560.

de Brito Marisa P., Valentina Carbone and Corinne Meunier Blanquart (2008) 'Towards a sustainable fashion retail supply chain in Europe: Organisation and performance', *International Journal of Production Economics*, 114(2), 534–553.

de Coverly, Edd, Pierre McDonagh, Lisa O'Malley and Maurice Patterson (2008), 'Hidden mountain: The social avoidance of waste', *Journal of Macromarketing*, 28(3), 289–301.

Desmond, John (2003), *Consuming Behaviour*, London: Palgrave.

Desmond, John, McDonagh, Pierre and S. O'Donohoe (2000), 'Counter-culture and consumer society', *Consumption, Markets & Culture*, 4(3), 241–280.

Dobscha, Susan., Prothero, Andrea and McDonagh, Pierre (2012), '(Re)-thinking distribution strategy: Principles from environmental sustainability', in N. Ozagler-Toulouse, L. Penaloza and L. Visconti (eds), *Marketing Management: A Cultural Perspective*, London: Routledge.

Fletcher, Kate (2008), *Sustainable Fashion and Textiles*, Earthscan: London.

Fletcher, Kate (2010), 'Slow fashion: An invitation for systems change', *Fashion Practice: The Journal of Design, Creative Process & the Fashion*, 2(2), 259–266.

Freeman, Esther (2012), 'Is the fashion industry blinding us with greenwashing', *Huffington Post*. Available online at http://www.huffingtonpost.co.uk/esther-freeman/is-the-fashion-industry- b_b_1633620.html (accessed 9 November 2013).

Fuller, Donald A., Jeff Allen and Mark Glaser (1996), 'Materials recycling and reverse channel networks: The public policy challenge," *Journal of Macromarketing*, 16(1), 52–72.

Giddens, A. (1986), *The Constitution of Society: Outline of the Theory of Structuration*, Berkeley, CA: University of California Press.

Giegerich, Andy (2013), 'Adidas, Nike take issue with Greenpeace's "Dextox Catwalk" Listing', *Sustainable Business Oregon*. Available online at http://www.bizjournals.com/portland/blog/sbo/2013/11/adidas-nike-take-issue-with-detox.html (accessed 9 November 2013).

GodeInlk, (2012), 'Is M&S's Shwopping a true revolution or a philanthropic effort doomed to fail?', *Triple Pundit*. Available online at http://www.triplepundit.com/2012/05/mss-shwopping-true-revolution-philantrophic-effort-doomed-fail/ (accessed 14 February 2014).

Goldstein, T.C. (2003), '*Nike v. Kasky* and the definition of "Commercial Speech",' *Cato Supreme Court Review*, 63–79. Available online at http://www.cato.org/pubs/scr2003/commercialspeech.pdf (accessed 12 November 2013).

Greenwashing Index (2012), 'M&S Shwop 2012 Campaign'. Available online at http://www.greenwashingindex.com/ms-shwop-2012-campaign/ (accessed 9 November 2013).

H&M (2013), Sustainability Policies. Available online at http://about.hm.com/en/About/Sustainability.html#cm-menu (accessed 9 November 2013).

Hamed, Brady (2013), 'The top 8 things that we learned from Marks and Spencer's 2013 Plan A Report', *Sustainable Brands*. Available online at http://www.sustainablebrands.com/news_and_views/communications/8-most-important- things-we-learned-marks-spencers-2013-plan-report (accessed 9 November 2013).

Kangun, Norman, Carlson, Les and Grove, Stephen. J. (1991), 'Environmental advertising claims: A preliminary investigation', *Journal of Public Policy & Marketing*, 10, 47–58.

Keithpp's Blog, (2012), 'M&S cynical exercise in greenwash'. Available online at http://keithpp.wordpress.com/2012/04/26/ms-cynical-exercise-in-greenwash/ (accessed 9 November 2013).

Kilbourne, W. E. (2004), 'Sustainable communication and the dominant social paradigm: Can they be integrated?' *Marketing Theory*, 4(3), 187–208.

Kilbourne, William E, McDonagh, Pierre and Prothero, Andrea (1997), 'Sustainable consumption and quality of life: A macromarketing challenge to the dominant social paradigm', *Journal of Macromarketing*, 17(1): 4–24.

Landler, Mark (1991), 'Suddenly, green marketers are seeing red flags', *Business Week*, (February 25), 74–76.

Lo, P. (2013), 'H&M responds slowly to Bangladesh factory collapse killing 1,100 by Puck Lo, CorpWatch Blog May 19th, 2013., Available online at http://www.corpwatch.org/article.php?id=15840 (accessed 7 November 2013).

McDonagh, P. (1995), 'Sustainable communication: Pipe dream for green advertisers or the new way for business to communicate?' in Bergadaa, M. (ed.), *Marketing Today and for the 21st Century*, Proceedings of the 24th European Marketing Academy Conference, Vol. 1, 731–750.

McDonagh, Pierre (1998), 'Towards a theory of sustainable communication in risk society: An empirical analysis relating issues of sustainability to marketing communications', *Journal of Marketing Management*, 14, 591–622.

Milbrath, L. (1984). *Environmentalists: Vanguards for a New Society*. Albany, NY: University of New York Press.

Miller, Ben (2012), 'Co-operative, M&S and Waitrose accused of greenwash over close links to "unethical" companies', *The Ecologist*. Available online at http://www.theecologist.org/News/news_analysis/1234402/cooperative_ms_and_waitrose_accused_of_greenwash_over_close_links_to_unethical_companies.html (accessed 9 November 2013).

Morley, N.J., Bartlett, C. and McGill I. (2009), *Maximising Reuse and Recycling of UK Clothing and Textiles*, A report to the Department for Environment, Food and Rural Affairs. Oakdene Hollins Ltd.

Nordic Initiative Clean & Ethical (2013) NICE Fashion. Available online at http://archive.today/cwCXB (accessed 9 November 2013).

O'Connor, (2001), 'Nike: Global compact violator: CorpWatch releases second in a series of exposés'. Available online at http://www.corpwatch.org/article.php?id=932 (accessed 7 November 2013).

Ozanne, Lucie K. and Julie L. Ozanne (2011), 'A child's right to play: The social construction of civic virtues in toy libraries', *Journal of Public Policy & Marketing*, 30(2), 263–276.

Packard, Vance (1960), *The Waste Makers*, New York: D. Mckay Co.

Peterson, Mark (2006), 'Focusing the future of macromarketing', *Journal of Macromarketing*, 26(2), 245–249.

Prothero, Andrea and James A. Fitchett (2000), 'Greening capitalism: Opportunities for a green commodity', *Journal of Macromarketing*, 20(1), 46–55.
Prothero, Andrea, McDonagh Pierre and Susan Dobscha (2010), 'Is green the new black? Reflections on a green commodity discourse', *Journal of Macromarketing*, 30(2), 147–159.
Sturdy, Andrew and Christopher Grey (forthcoming), 'Exploring alternatives: Beneath and beyond organizational change management', *Journal of Organisational Change Management*.
Sustainable Apparel Coalition (2013), http://www.apparelcoalition.org, (accessed 9 November 2013).
TodayEco (2013), 'H&M conscious leaves workers "unconscious" says anti-sweathshop group'. Available online at http://todayeco.com/fashion/pages/12301010-h-and-m-conscious-leaves-workers-unconscious-says-anti (accessed 9 November 2013).
Volokh, E. (2002), 'Nike and the free-speech knot', *Wall Street Journal*, June 30, at A16. Available online at http://www2.law.ucla.edu/volokh/nike.htm (accessed 12 November 2013).

Sustainability landing sites

Nike – http://nikeinc.com/pages/responsibility
H&M – http://about.hm.com/en/About/sustainability.html
Marks and Spencer – http://corporate.marksandspencer.com/plan-a

PART IV
Preventing waste

11

UPCYCLING OF PRE-CONSUMER WASTE

Opportunities and barriers in the furniture and clothing industries

Daniel Hjelmgren, Nicklas Salomonson and Karin M. Ekström

Introduction

The consumption of clothes and textiles in society has increased over time parallel to the development of the consumer society. The net inflow (imports plus domestic production − exports) of textiles to Sweden in 2008 was 131,800 tonnes, which corresponds to almost 15 kg per person (Carlsson *et al.* 2011). The net inflow of textiles increased by nearly 40 per cent during the period 2000–2009 (Carlsson *et al.* 2011). These changes are also noticeable in many other countries. Studies in the UK show that the volume of sold clothing increased by 60 per cent between 1995 and 2005 (Morley *et al.* 2006). One reason for the increased consumption is an increased emphasis on fashion in society and fashion cycles that change faster every year.

Increased consumption has also led to an increase of waste. In Sweden, it is estimated that an average consumer throws away about 8 kg textiles per person per year (Carlsson *et al.* 2011). In the UK, the figure is about 30 kg each year (Allwood *et al.* 2006). This is a problematic development and in many parts of the world textiles have become an increasing landfill problem. Landfills pose environmental concerns by polluting air and water, and represent a waste of resources (Wang 2010). In Sweden, the solution is instead to burn the textile waste in large incinerators to create energy. This, however, is also a waste of resources, since a lot of natural resources are used in the production of textiles, and the energy gained by burning textiles is much less than what is needed to produce it. Approximately 7,000 to 29,000 litres of water and 0.3 to 1 kg of oil are used to produce a kilogram of cotton; the exact amount depends on where the cotton is produced (Fletcher 2008). In addition, there are pesticides that damage the environment. The production of alternative fibre materials, such as polyester and nylon, also has an impact on the environment. Producing 1 kg of polyester uses 109 MJ of energy and 1 kg of nylon uses 150 MJ of energy (Fletcher 2008). Hence, energy would be saved if textile waste was recycled instead of being thrown away and burned in large incinerators.

Recycling could be one way to reduce the amount of waste of textiles. It involves reuse, remanufacturing or reprocessing. Examples of reuse are second-hand clothes or reuse of waste textiles as fillings in mattresses and furniture, or as sound insulation in cars. Reprocessing occurs when materials are recycled in an industrial context where textiles are unravelled and re-spun into new fibres. An example is Teijin, a Japanese polyester supplier, which allows unwanted polyester garments to be recycled into PET polymer used in the remanufacturing of new garments. Upcycling is reuse where waste is transformed into new products of either equal or better quality or a higher environmental value (Smusiak 2010). An example of this is when old curtains are turned into garments or when an old pair of jeans is made into a bag.

There is increased interest among authorities, retailers, charity organisations, recycling companies, consumer organisations etc. to deal with this societal problem (Ekström and Salomonson 2014). Focus has mainly been on post-consumer waste, but lately an increased interest in pre-consumer waste has been noticeable. Pre-consumer waste is the waste that results from the process of manufacturing and distributing a product for sale and consumption. It is, for example, caused by mistakes in design, wrong colours, machine problems, fabric faults, trimming scarp and faster fashion cycles. Braungart and McDonough (2002) claim that final products overall contain only 5 per cent of the total raw materials required for making and delivering them, i.e. they only constitute the 'tip of the iceberg' of the total waste created. This signals the need to also consider what to do with pre-consumer waste. Pre-consumer waste is particularly a problem in the textile industry that depends on fashion. Most of the textile waste from production in Sweden is currently being burned and this is clearly a waste of resources.

An example of an initiative to reduce pre-consumer waste is Studio Re:design, an EU-funded project that was established in 2012 by the region of Västra Götaland (VGR) in Western Sweden. The aim of the project has been to identify new ways to make innovative products from pre-consumer textile waste, so called *upcycling*, and to use it on a large scale in the Swedish furniture and clothing industries. The purpose of this chapter is to provide insights about opportunities and structural barriers for large-scale upcycling in the Swedish furniture and clothing industries, as identified in the Studio Re:design project. We adopt an Industrial Network Approach (INA) and focus on two central concepts: resource combination and resource embeddedness. Before the results and conclusions are presented, we discuss the theoretical framework and method for studying Studio Re:design.

Theoretical framework

This chapter concerns the development of new products through new resource combinations, and how certain features of these products create opportunities through different types of potential value for consumers. Zeithaml (1988) stated that perceived value is consumers' overall assessment of the utility of a product or service based on their perceptions of what is received versus what is given up. The early definitions primarily focused on the trade-off decision between quality and price (Monroe 1990), which has since been argued to be too simplistic and incomplete (Woodruff 1997). Other important factors are social identity (Vickers and Renand 2003); uniqueness (Ruvio 2008); hedonistic attributes such as colour, fashion and style; functional attributes such as durability and comfort (Sondhi and Singhvi 2006); and brand (Carrigan and Attalla 2001). Furthermore, consumers purchasing clothing are increasingly considering ethical attributes (Kim and Damhorst 1998).

Another important issue in the chapter is to analyse opportunities and barriers for using textile waste due to resource embeddedness. According to the INA, resources can be divided into tangible and intangible resources (see e.g. Håkansson 1987). While tangible resources include physical assets, such as production equipment, components and material, intangible resources include knowledge, skills and routines. Both tangible and intangible resources are embedded in a network of other resource units (Håkansson and Snehota 1995; Baraldi and Strömsten 2008; Gadde et al. 2012; Hjelmgren and Dubois 2013). Contemporary embeddedness research is largely influenced by Granovetter's research (1985), where he argues that organisations' decision-making is largely affected by the networks of on-going interpersonal relationships in which organisations are embedded. In addition to this social dimension of embeddedness, there is a technical dimension of embeddedness. An example of research concerning the latter is Hughes (1987), who argues that a technological system consists of different artefacts that interact and are adapted to each other to work as an integrated whole. If an artefact is taken away from the system or if its characteristics are changed, additional changes of other artefacts in the system may need to be made. Re-establishing complementarities that have been disrupted usually calls for extra time and money (Dhebar 1995). In addition, the change of a resource may affect its value in other resource combinations (Wedin and Johansson 2000; Hjelmgren 2011). Hence, when combining resources, companies need to consider the network of other resources in which they are embedded.

The large number of connections that commonly exist between different resources makes it difficult to foresee the total effects of a particular change (Baraldi 2003; Harrison and Håkansson 2006). The principle of non-proportional growth says that 'when any structure grows, the proportions of its parts and of its significant variables cannot remain constant. That is to say, it is impossible to reproduce all the characteristics of a structure in a scale model of different size (Boulding 1953: 335). Therefore, 'subsequent adjustments of a change are not necessarily proportional to the change itself' (Dubois et al. 2002: 60). Uncertainty regarding the consequences of a particular change, results in companies often being less willing to make changes (Hjelmgren 2011).

For structuring analysis of resource combination and embeddedness, Håkansson and Waluszewski (2002) separate technical resource units into products and production facilities. Products involve both single physical items and systems of items, including additional services such as support and maintenance. The set of physical items and additional services gives each product specific features affecting its potential value for different users. Production facilities involve different types of equipment, premises, skills and routines.

Method

The study is based on a single case study (see e.g. Yin 2003) of an EU-funded project called Studio Re:design where three designers developed clothing and furniture prototypes based on pre-consumer waste from nine textile manufacturers in Sweden. Their work resulted in a collection of about 50 prototypes that were shown during the Stockholm Furniture Fair in February 2013. As researchers, we had the opportunity to follow the project from the very start by conducting semi-structured face-to-face interviews and by observing and taking part in different activities during the project. Examples of activities were the inauguration of the Studio Re:design project in September 2012 and the presentation of prototypes from the project in March 2014. Interviews were made with the project leader, the assistant project leader and the two business developers who were involved in the project. One interview was conducted with the three designers employed in the project and nine interviews with different companies who either supplied surplus textile to be used in the project or were potential customers of the project's end results. The interviews were conducted between autumn 2012 and spring 2013. Each interview lasted between 25 minutes (companies) to 2 hours (project leader), and were tape recorded and transcribed shortly after. The interviews focused on the opportunities and barriers for using pre-consumer textile waste in the production of clothing and furniture. In addition, nine short interviews of participating companies at the Stockholm Furniture Fair in February 2013 (approximately 5–10 minutes each) were conducted to find out how they perceived the clothing and furniture prototypes from the Studio Re:design project that were displayed at the fair. Information from secondary data sources was also collected during the project, such as the project website (epi.vgregion.se/studioredesign/), company websites, project meeting protocols and a brochure developed for the Stockholm Furniture Fair. The analysis of the interview transcriptions, the observations and the secondary data was conducted in accordance with traditional qualitative techniques, that is, an interactive process of sorting and categorising the results.

Results

The Studio Re:design project consists of two parts. The first part concerns the development of clothing prototypes and the second part concerns development of furniture prototypes. In the first part of the project (development of clothing prototypes), three companies were mobilised. Their different product areas and contributions to the Re:design project are presented in Table 11.1.

TABLE 11.1 The companies' product areas and contributions to the clothing project

Company	Product areas	Material provided
Abecita	Manufactures and sells bras, nursing bras, sports bras, ladies briefs, bodies, bikinis, swimwear and nightwear	Corset laces
Eton	Manufactures and sells high-brand shirts made of wrinkle-free cotton woven in Switzerland exclusively for Eton	Fabrics scrap from its production of shirts
Nudie Jeans	Manufactures and sells raw denim jeans, jackets, shirts and accessories	A variety of different unsold goods

In the second part of the project (development of furniture prototypes), seven companies were mobilised. Their different product areas and contributions to the Re:design project are presented in Table 11.2.

The resource combination performed in the furniture project and the clothing project is analysed below. First, opportunities are analysed in terms of how different product features with potential value for end customers are developed when certain resources are combined. The analysis focuses on two specific prototypes (one clothing prototype and one furniture prototype) that were developed in the Re:design project. Second, barriers are analysed in terms of lost/reduced value in other parts of the resource network.

Opportunities

The first prototype analysed is a jacket (see Figure 11.1). The jacket is a combination of waste material from Ludvig Svensson and the three designers' knowledge and skills in designing clothes (see Table 11.3). The product has three features with potential value for end customers: it is 1) sustainable, 2) unique, and 3) has a high-tech feeling. The first feature, sustainable, is based on waste material being used. The second feature, uniqueness, is based on the fact that the availability and sorts of waste material may vary widely, and thus make each product more or less unique. The third feature, a high-tech feeling, emerges when combining shirt fabric from Eton and greenhouse fabric from Ludvig Svensson. One material feature that was used in the prototype was the greenhouse fabric's transparency.

The prototype developed in the furniture project is a sofa (see Figure 11.2). The sofa is a combination of furniture frames from Keystone, waste material from FOV, and the three designers' knowledge and skills in designing clothes (see Table 11.4). This combination resulted in four product features with potential value for end customers: it is 1) sustainable, 2) unique, 3) has a high-tech feeling, and 4) is a new form of expression.

Just as for the jacket, the first and second features are due to waste material being used. The third and fourth features, a high-tech feeling and a new form of

TABLE 11.2 The companies' product areas and contributions to the furniture project

Company	Product areas	Material provided	Production facility provided
Ludvig Svensson	Textile for professional interior: terry, textiles and technical fabrics for climate controlling in greenhouses	Technical fabrics used in greenhouses	
Olle Winter	Contract worker sewing textile interiors for homes and public spaces: upholstery fabrics, curtain fabrics, ready-made-curtains, cushion covers, bed covers, bed skirts, curtain tie-backs, laces and inner cushions		Production facility involving sewing machines with the ability to sew a large range of different patterns
Keystone	Stylish furniture for public areas, e.g. hotels, restaurants, conferences, hospitals, schools and offices	Furniture frames	
FOV	Develop, manufacture and market technical woven textile and functional garment fabrics used by suppliers to the automotive industry as well as by producers of sportswear garments	Fabric used in airbags	
NUD	Manufactures lamps with a simple design consisting of just a socket and a cable in full colour	Lamp cords in different colours	
Elmo Sweden	A tannery which produces and sells leather to the furniture and the automotive industries	Old samples of leather	
Acqwool	Develops and manufactures sound absorbers made from 100 per cent wool	Waste materials arising when the machines are switched from producing one type of material to another	

FIGURE 11.1 The jacket prototype. Photographer: Jan Töve

TABLE 11.3 Resource combination in the clothes project – the jacket prototype

Product features	Products	Production facilities
Sustainable Unique High-tech feeling	Waste material, fabric used in shirts – Eton Waste material, technical textiles used in greenhouses – Ludvig Svensson	The designers experience from the fashion industry and sewing skills

FIGURE 11.2 The sofa prototype. Photographer: Jan Töve

TABLE 11.4 Resource combination in the furniture project – the sofa prototype

Product features	Products	Production facilities
Sustainable Unique High-tech feeling New form of expression	Furniture frames – Keystone Waste material, technical textiles used in airbags – FOV	The designers experience from the fashion industry and the type of seams used there

expression, were created by the combination of furniture frames from Keystone, designers with experience from the fashion industry (and the type of seams used there) and technical textiles (airbag fabric) from FOV. Material features that were used were the air bags' pink colour and high-tech association. In addition to these features, the designers' earlier knowledge and experience from different type of seams were used. For a long time the furniture business has been dominated by designers with a background in architecture, which has resulted in a large supply of austere furniture often in dark and dull colours.

Besides new ideas of different combinations of products and production facilities, the Re:design project also resulted in an adjusted design process fostering designers' creativity. Instead of starting right from the very beginning with a few restrictions for what can be made, the designers were forced to start from a limited set of (waste) materials focusing on how to use and combine these materials in feasible and novel ways. Due to this modified design process, the designers came up with new ideas that otherwise would not have appeared. According to the designers, they would not normally have designed a jacket where the pattern of the lining is visible from the outside.

Barriers

An important barrier for upcycling of textile waste is the shortage of suitable production facilities, since a large part of the Swedish fashion industry's manufacturing is situated abroad. To send the clothes back and forth to Turkey, for example, would result in transportation costs that tripled compared to new production. Moreover, it would not be environmentally sustainable. Furthermore, clothing production based on waste material requires large raw material inventories so that it can be produced and transported efficiently. In addition, clothing production facilities close to Studio Re:design, for example Eton's facilities, are either too specialised to be used for purposes other than intended or on too small scale.

The sale of clothing made of waste materials is difficult since distributors have established routines for purchasing new collections and commonly place exact orders regarding quantity, colour, size, form etc. Since production based on waste material is dependent on the availability of different waste material, it requires *production on speculation*, i.e. that each product is produced before the customer makes any decision to buy it. It not only requires large spaces for inventory, but may also result in more capital being tied up in inventory, as well as increasing the

TABLE 11.5 Costs and lost/reduced values that might appear in the resource network

Products	Production facilities
Costs due to capital tied in inventory	Transportation costs and lost/reduced value that is created from an environmental perspective due to transportation of finished goods, or high production costs due to small scale
Costs for producing clothes which are not in demand	Costs of inventories
Reduced perceived value of products using the same brand name as the product made of waste material	When using a highly specialised production facility, production of products made of waste material has a significant negative impact on the utilisation of the facility
	The need for distributors to change their purchasing routines

risks of producing clothes for which there is no demand. Finally, selling clothes made of waste material is hindered by some clothing companies' cautiousness about their brands. In this study, one of the clothing companies was reluctant to use 'unknown' materials in its products. Moreover, it was not possible to use the company's own highly specialised production facility in Sweden, and the production would therefore need to be located at production facilities unknown to the company. Table 11.5 summarises the barriers for upcycling pre-consumer textile waste due to costs or lost/reduced value that might appear in the resource network.

The barriers for upcycling of textile waste appeared to be less in the furniture industry. As mentioned above, the project managed to mobilise Olle Winter, a company that has a small, but flexible production facility that facilitates shorter production series without reducing the utilisation of certain production equipment due to increased set up times. Furthermore, the possibility of shorter production series reduces the need for a large inventory of finished products. However, due to varying access to different types of waste material, customers cannot count on the possibility of choosing between a large range of combinations of colour and material. Neither can they count on purchasing a large amount of furniture in the same material and colour. This is especially not possible if ordered goods need to be delivered within a relatively short timeframe. In such situations, using waste material requires a large inventory of finished products.

Conclusions

The purpose of this chapter has been to identify the opportunities and structural barriers for more large-scale upcycling in the Swedish furniture and clothing industries. Opportunities are created by the potential value for customers when new combinations of resources are made possible. A result from the study is that products produced with waste material create value not only for environmentally conscious customers, but also for customers who are interested in unique products and new, exciting design. The opportunities for exciting design were created by, for example, using material that originally was not intended to be used in furniture and clothing. New innovative prototypes were developed not only as a result of the interaction between different material resources, but also as the result of the interaction between material and immaterial resources. For example, the designers' knowledge about different types of seams that are used in clothing was used to give the sofa a unique form of expression. Most importantly, an adjusted design process fostered the designers' creativity.

Barriers to the production and sale of re-designed furniture and clothing are caused by the resources' embeddedness in the network of other resources and the extent to which a new resource combination affects other parts of the resource network. The study showed, for example, how the possibilities for production of clothing made of waste material were influenced by the fact that the production facilities nearby are highly specialised, i.e. could not be used without large adjustments and various negative effects on the utilisation of certain production equipment. Barriers were also caused by costs due to the need to adjust to surrounding structures, for example more space for the raw material inventory and more production on speculation. In order not to let the ready-made inventory become too large, it is necessary that the distributors of clothes and furniture adjust their purchase patterns to the varying supply of waste material.

When comparing the furniture project and the clothing project, it is clear that the furniture project had an advantage in having companies nearby with flexible production facilities. This makes it possible to produce shorter production series of re-designed furniture without the production cost becoming higher than for comparable products on the market. Furthermore, an additional interactive effect between incoming resources in the furniture project was developed when classic furniture frames were combined with the designers' knowledge about clothing design resulting in new forms of expression.

This chapter is about the opportunities and structural barriers for more large-scale upcycling in the Swedish furniture and clothing industries. Although there are similarities with manufacturing of clothing and furniture in other countries, at least countries in the Western world, this study cannot with certainty say anything about opportunities and barriers in general. In addition, the availability of waste material from production can differ between countries. Future research could therefore focus on opportunities and barriers for upcycling of clothing and furniture in other countries. Is Sweden unique or is a similar pattern being found elsewhere? The potential of upcycling definitely deserves more attention in a society facing problems of increasing waste for both pre-consumer and post-consumer waste. Creativity is here a vital component. Upcycling is, however, not the only solution to a problem based on overproduction and overconsumption. Changing consumption patterns and a fashion industry that seriously considers sustainability are not only desirable, but necessary in a society dealing with an increase of textile waste.

References

Allwood, Julian M., Sören E. Laursen, Cecilia M. de Rodríguez and Nancy M.P. Bocken (2006), 'Well dressed? The present and future sustainability of clothing and textiles in the United Kingdom', University of Cambridge, Institute for Manufacturing, Mill Lane, Cambridge CB2 1RX, UK. Available online at http://www.ifm.eng.cam.ac.uk/news/cambridge-report-lays-out-options-for-an-environmentally-sustainable-fashion-industry/#.U-3T7od0w8w (accessed 15 August 2014).

Baraldi, Enrico (2003), *When Information Technology Faces Resource Interaction: Using IT Tools to Handle Products at IKEA and Edsbyn*, Ph.D. thesis, Department of Business Studies, Uppsala University, Uppsala.

Baraldi, Enrico and Torkel Strömsten (2008), 'Configurations and control of resource interfaces in industrial networks', *Advances in Business Marketing and Purchasing*, 14, 251–316.

Boulding, Kenneth (1953), 'Toward a general theory of growth', *The Canadian Journal of Economics and Political Science*, 19, 326–40.

Braungart, Michael and William McDonough (2002), *Cradle to Cradle: Remaking the Way We Make Things*, New York: North Point Press.

Carlsson, Annika, Kristian Hemström, Per Edborg, Åsa Stenmarck and Louise Sörme (2011), 'Kartläggning av mängder och flöden av textilavfall. SMED på uppdrag av Naturvårdsverket' [Mapping the amount and the flow of textile waste. SMED on behalf of the Swedish Environmental Protection Agency], Report No. 46. Norrköping, Sweden: Sveriges Meteorologiska och Hydrologiska Institut [Swedish Meteorological and Hydrological Institute].

Carrigan, Marylyn and Ahmad Attalla (2001), 'The myth of the ethical consumer: Do ethics matter in purchase behavior', *Journal of Consumer Marketing*, 18, 560–77.

Dhebar, Anirudh S. (1995), 'Complementarity, compatibility, and product change: Breaking with the past?' *Journal of Product Innovation Management*, 12, 136–52.

Dubois, Anna, Daniel Hjelmgren and Håkan Håkansson (2002), 'Technical development in networks: The importance of third parties', *Sinergie*, 58, 45–64.

Ekström, Karin. M. and Nicklas Salomonson (2014), 'Reuse and recycling of clothing and textiles: A network approach,' *Journal of Macromarketing*, 34(3), 383–99.

Fletcher, Kate. (2008), *Sustainable Fashion & Textiles*, Design Journeys, London: Earthscan.

Gadde, Lars-Erik, Daniel Hjelmgren and Fredrik Skarp (2012), 'Interactive resource development in new business relationships', *Journal of Business Research*, 65, 201–17.

Granovetter Mark (1985), 'Economics action and social structure: The problem of embeddedness', *American Journal of Sociology*, 91(3), 481–510.

Håkansson, Håkan (1987), *Industrial Technological Development: A Network Approach*, London: Croom Helm.

Håkansson, Håkan and Ivan Snehota (1995), *Developing Relationships in Business Networks*, London: Routledge.

Håkansson, Håkan and Alexandra Waluszewski (2002), *Managing Technological Development*, London: Routledge.

Harrison, Debbie and Håkan Håkansson (2006), 'Activation in resource networks: A comparative study of ports', *Journal of Business & Industrial Marketing*, 21, 231–38.

Hjelmgren, Daniel (2011), 'Combining resources and limiting the change boundary: The case of an ERP system implementation', *Innovative Marketing*, 7(2), 8–19.

Hjelmgren, Daniel and Anna Dubois (2013), 'Organising the interplay between exploitation and exploration: The case of interactive development of an information system', *Industrial Marketing Management*, 42(1), 96–106.

Hjelmgren, Daniel and Eva Gustafsson (2013), *Textilreturen i Ullared: ett experiment om återvinning [The Textile Return in Ullared: An Experiment About Recycling]*, Vetenskap för profession [Science for the Professions], No. 25, Borås: University of Borås.

Hughes, Thomas P. (1987), 'The evolution of large technological systems', in Wiebe, E. Bijker, Thomas P. Hughes, Trevor Pinch and Deborah G. Douglas (eds), *The Social Construction of Technological Systems*, Cambridge, MA: MIT Press, 51–82.

Kim, Hye-Shin and Mary Lynn Damhorst (1998), 'Environmental concern and apparel consumption', *Clothing and Textiles Research Journal*, 16(3), 126–33.

Monroe, Kent B. (1990), *Pricing: Marketing Profitable Decisions*, New York: McGraw-Hill.

Ruvio, Ayalla (2008), 'Unique like everybody else? The dual role of consumer's need for uniqueness', *Psychology & Marketing*, 25, 444–64.

Smusiak, Cara (2010), 'What is upcycling?' *Naturally savvy*. Available online at http://www.naturallysavvy.com/naturally-green-faq/what-is-upcycling (accessed 15 August 2014).

Sondhi, Neena and S. R. Singhvi (2006), 'Gender influencing garment purchase: An empirical analysis', *Global Business Review*, 7(1), 57–75.

Vickers, Jonathan S. and Renand, Franck (2003), 'The marketing of luxury goods: An exploratory study – three conceptual dimensions', *Market Review*, 4, 459–78.

Wang, Youjiang (2010), 'Fiber and textile waste utilization', *Waste and Biomass Valorization*, 1, 135–43.

Wedin, Torkel and Martin Johanson (2000), 'Value creation in industrial networks', paper presented at the 16th Annual IMP Conference, University of Bath, Bath.

Woodruff, Robert B. (1997), 'Customer value: The next source of competitive advantage', *Journal of the Academy of Marketing Science*, 25(2), 139–53.

Yin, Robert K. (2003), *Case Study Research: Design and Methods*, London: Sage.

Zeithaml, Valarie A. (1988), 'Consumer perceptions of price, quality, and value: A means-end model and synthesis of evidence', *Journal of Marketing*, 52(3), 2–22.

12
POST-OWNERSHIP SUSTAINABILITY

Russell Belk

The end of ownership?

It is estimated that the average lifetime use of an electric drill of the sort that many consumers own is between 6 and 20 minutes (Steffen 2007). Cars are used about one to two hours a day; the rest of the time they depreciate in value, accumulate insurance and maintenance costs, and take up increasingly limited and often quite expensive parking space. When we buy or download a season's worth of a television series, we seldom watch the shows more than once. Arnold and Lang (2007) found that in southern California, the majority of the large (2–3 car) garages could no longer fit even a single car because they had been transformed into storage areas. There must be a better way of making use of these and the many other things we seldom use that largely add to the clutter of our already crammed homes.

A better way may lie in seeking to access rather than own many of the things we now acquire as personal possessions. Consider a few of the indications that we are moving in this direction:

- Driving by young people decreased by 23 per cent between 2001 and 2009 (Rosenthal 2013). Part of the reason is car-sharing and bike-sharing programmes that are proliferating in North America and Europe.
- The amount of driving is inversely related to the availability of the Internet. Taking mass transit means that young commuters can be online.
- Fewer young people are getting a driver's licence at all. Symbols of freedom have changed. As one young woman put it, 'When I try to imagine my dream car, I draw a blank. Then I reach for my phone' (Brooks 2013).
- Uber, a ride-sharing site has received $258 million in venture capital from Google Ventures (Wohlsen 2013).

- On any given night, 40,000 people rent accommodation from Airbnb rather than a hotel. The peer-to-peer (P2P) home rental organization offers 250,000 rooms in 30,000 cities in 192 countries (*The Economist* 2013).
- More than 3 million people from 235 countries have used the free home-sharing and local acquaintance-making application, CouchSurfing, and 2.2 million bike-sharing trips are taken each month (Sacks 2011).
- Sharing and rental economies generated $3.5 billion in revenue in 2013, a figure forecast to grow to $110 billion in 5 years (Boesler 2013).

Thanks in part to the economic recession that began in 2009 as well as the availability of Web 2.0 interactivity to facilitate knowing where, how, and why to access an increasing array of goods, these and other sharing and collaborative consumption options are becoming increasingly popular. And they just may be ushering in a new era of more sustainable consumption both socially and ecologically.

There have been a number of recent books and papers about the new 'sharing economy', using labels that include 'access-based consumption' (Bardhi and Eckhardt 2012), 'co-creation' (Prahalad and Ramaswamy 2004; Lanier and Schau 2007), 'collaborative consumption' (Botsman and Rogers 2010), 'commercial sharing systems' (Lamberton and Rose 2012), 'consumer participation' (Fitzsimmons 1985), 'co-production' (Humphreys and Grayson 2008), 'the mesh' (Gansky 2010), 'online volunteering' (Postigo 2003) 'product-service systems' (Mont 2002), 'prosumption' (Ritzer and Jurgenson 2010; Toffler 1980), and sharing (Belk 2010). There is a correspondingly large number of for-profit and non-profit businesses that participating in the 'sharing economy' (e.g. Lessig 2008; Sacks 2011), including Airbnb, Facebook, Flickr, Freecycle, Twitter, Wikipedia, YouTube, and Zipcar. The Internet in general acts as a huge pool of shared content for all to access. What these various labels and corresponding parts of the sharing economy have in common is their reliance on the Internet to facilitate sharing arrangements and their provision of temporary access to consumer goods and services. In this chapter I focus on a comparison of sharing and collaborative consumption, examine the extent to which various parts of the 'sharing economy' truly involve sharing, and assess the prospects for more sustainable consumption. I draw on prior research by others as well as on my own prior conceptual (Belk 2007, 2010) and empirical (Belk and Llamas 2012) work.

Theoretical considerations

Belk (2010) contrasts the prototypes of sharing (mothering and the pooling and allocation of household resources) with the prototypes of gift giving (the exchange by Della and Jim in the O. Henry story 'The Gift of the Magi') and of marketplace exchange (buying bread at a store for money). In an earlier paper, sharing was defined as 'the act and process of distributing what is ours to others for their use and/or the act and process of receiving or taking something from others for our use' (Belk 2007: 126). Benkler (2004) simply treats sharing as 'nonreciprocal pro-social behaviour'.

'Sharing in' occurs when the sharing makes the recipient a part of a pseudo-family and our aggregate extended self (Belk 1988, 2010, 2013; Ingold 1986). When the sharing involves a one-time interaction between strangers or when it involves dividing something up so that those in the transaction can go their separate ways it is labelled 'sharing out'. Thus, besides involving varying levels of intimacy, sharing in and sharing out (as well as shades between them) vary in their expectations of continued bonds between parties participating in the sharing. Both sharing in and sharing out differ from gift giving and marketplace exchange because they do not transfer ownership, are not reciprocal, and do not produce a debt.

Borrowing and lending are borderline cases between sharing and commodity exchange, although 'borrowing' is sometimes a euphemism for requested sharing, as in 'Can I borrow a paper clip?' Nevertheless, even this small act of sharing may facilitate bonding and intimacy. That is, it may be an act of sharing in, whereas picking up a hitchhiker is more apt to be a case of sharing out.

We can share intangibles like ideas, values, and time (Belk 2010). But mere coincidences such as sharing a birthday or a language are not volitional acts of sharing. 'Demand sharing' and 'open sharing' are two types of sharing that frequently occur within families. Demand sharing requests generally must be met and occur when hunters in hunter–gatherer societies are obliged to share their kills with the community or when our children ask to be fed. We would also probably not begrudge a request for directions or the time of day. Open sharing means that our family members or house-guests can freely access certain resources such as food or use of the bathroom without having to ask. Guests may only achieve this level of intimacy over a history of sharing experiences with us.

Why now?

Ownership concepts have a long history. Although Plato believed that communal ownership is the basis for the good society, Aristotle believed that private property made for good citizens. Neither was troubled by slave ownership, however. John Locke's theory of labour value maintained that if a person planted a fruit tree and cared for it, he or she was entitled to what were literally the fruits of his or her labour. But Pierre-Joseph Proudhon proclaimed that property is theft and Karl Marx's theory of labour value held that capitalists unjustly appropriate worker labour and alienate them from the fruits of their labour. Karl Polyani (1944) maintained that only with the enclosure of the commons that started in about 1790 did land, labour, and capital become 'fictitious commodities'. Although there have been numerous attempts to foster alternative communal consumption throughout history, it is only now with the rise of the digital age that evidence of large-scale shared consumption practices have begun to emerge. Because there is such a diverse array of practices now invoking the concept of sharing (Wittel 2011), it is necessary to first make some distinctions between different types of Internet-facilitated sharing.

The economic downturn following the US mortgage meltdown of 2008, the subsequent bank crisis, and their global ramifications were all factors that have precipitated increased attention to sharing; the Internet and especially Web 2.0 have played a major role in facilitating new ways of sharing, leading to what Grassmuck (2012) calls 'the sharing turn'. Peer-to-peer (P2P) file-sharing of music and films started with Napster and has continued with various Bit Torrent successors (Giesler 2006; Hennig-Thurau et al. 2007). Online marketplaces like eBay have introduced buyer and seller ratings of each other in what has been called a 'reputational economy' (Masum and Tovey 2011; Sacks 2011; Solove 2007). While eBay and other online marketplaces do not involve sharing, the reputational systems they have introduced have also facilitated establishing trust within sharing systems like CouchSurfing and in collaborative consumption systems like Airbnb. However, online file distribution has also prompted the music and film industries to incorporate digital rights management (DRM) into their products, to prosecute downloaders, and to seed the Internet with corrupt files to foul downloading. The once idealistic digital dream that information wants to be free, has been partly suppressed by an increase in Intellectual Property Rights (IPR) legislation (Giesler 2008). Nevertheless, the 'war on sharing' has not been entirely successful and many consumers still download much of their music and many of their movies for free (Aigrain 2012). One downloading site, The Pirate Bay, has even had advocates elected to the European Parliament. In a battle of labelling between piracy versus sharing, The Pirate Bay has re-appropriated the term piracy and reclaimed it as both rebellious and virtuous.

There are legal online sites allowing downloads and streaming of music, television programmes, and movies for a price, including the new Napster, iTunes, Rhapsody, Pandora, Netflix, and Spotify. But for young people, illegal downloading remains prevalent. A CBS (2009) survey found that among 18- to 29-year-olds in the US, 69 per cent believe that it is okay to share music online either sometimes or always.

Although file-sharing or sharing of 'information goods' (Galbreth et al. 2012) might seem to be the purest form of sharing, Bit Torrent-based applications count user downloads and uploads to ensure that there is the balance needed to keep the system working (Aigrain 2012). Slater (2000) found that even among those trading sex pictures freely downloaded from the Internet, a balance rule was enforced. Such models appear to be more like barter, a form of non-market exchange that is not sharing per se (Belk 2010). It is considered below as a form of collaborative consumption.

Although file-sharing has received the greatest amount of media attention, the most prevalent form of online sharing involves sharing information. This includes contributions to blogs; wikis like Wikipedia; open-source software like Linux; photo- and video-sharing sites like Flickr and YouTube; online forums of many sorts; Twitter tweets; ratings on sites like TripAdvisor, AngiesList, and Amazon; and social media like Facebook, Pinterest, SnapChat, and Instagram. YouTube asks us 'What do you have to share today?' (John forthcoming). And although heavy contributors can receive some compensation or free merchandise, most of these contributions to the vast cornucopia of the Internet are not compensated (e.g. Hemetsberger 2012; Reagle and Lessig 2010).

The Internet also facilitates the sharing or collaborative consumption of material goods (Belk and Llamas 2012). Unlike eBay, craigslist, and Kijiji, which primarily offer goods for sale, sites like Freecycle and Really Really Free Market (e.g. Arsel and Dobscha 2011; McCartney 2012; Willer *et al.* 2012) offer, as their names imply, free goods and services to anyone without reciprocal obligations. Although these still often involve a transfer of ownership, they also include sharing skills, services, and activities like hugs, songs, and massages. There are also many true sharing exchanges including tool libraries and toy libraries that use the Internet to promote and reserve what is available (e.g. Ozanne and Ballantine 2010). Sweden also has clothing libraries. Just as you can check out a free book at a book library, you can check out tools, toys, and clothing at these libraries. Similarly, the Sharehood (see www.thesharehood.org), was started in the Northcote neighbourhood of Melbourne by Michael Green when he needed to do some laundry. He knew that lying between him and the nearest Laundromat there were many homes with idle washing machines. So he started the Sharehood to bring together people in need of various tools and appliances and those who are willing to share them. He began this online sharing service for utilitarian reasons, but found that the biggest ultimate benefit was meeting his neighbours. These neighbours learned that by sharing tools, appliances, and home care implements, they created a real sense of community. The concept has grown and spread to many other cities around the world.

Collaborative consumption

Botsman and Rogers' (2010, p. xv) definition of collaborative consumption as including 'traditional sharing, bartering, lending, trading, renting, gifting, and swapping' is too broad. It also mixes marketplace exchange, gift giving, and sharing. They are more accurate in giving the example of how Joe Gebbia, Brian Chesky, and Nathan Blecharcyzk conceptualized Airbnb.com:

> On a whiteboard in their apartment they drew a spectrum. On one side they wrote 'hotels' and on the other they scribbled rental listings such as craigslist, youth hostels, and nonmonetary travel exchanges such as Couch-Surfing that help people travel by creating a network of 'couches' available to sleep on for free.
>
> *(Botsman and Rogers 2010, p. x)*

They chose a middle ground, which includes sharing a home like CouchSurfing, yet in which the owners charge visitors a fee like a hotel. The monetary exchange keeps this from being a case of true sharing, like CouchSurfing. Yet this is a non-market form of exchange involving short-term rental (Belk forthcoming).

Collaborative consumption involves people intentionally coordinating their acquisition and use of a resource for a fee or other compensation. It includes bartering, trading, and swapping, all of which involve non-monetary compensation. But it excludes true sharing activities like those of CouchSurfers who are prohibited from offering or accepting compensation. It also excludes gift giving since giving a gift involves a transfer of ownership. For example, if I give my old bicycle to the Salvation Army, this is a gift because it involves a transfer of ownership. If I instead tell neighbours to feel free to use it whenever they wish, this is sharing. And if I offer to rent it out for a daily fee, that would be collaborative consumption. Thus, collaborative consumption occupies a position between sharing and marketplace exchange, and has elements of both. Like Zipcar.com (Bardhi and Eckhardt 2012), most .com 'sharing' organizations involve offering collaborative consumption opportunities. I (Belk forthcoming) have called such transactions 'pseudo-sharing' in that they often invoke a vocabulary of sharing (e.g., 'car sharing'), but actually involve short-term rental.

Bardhi and Eckhardt (2012) use the term 'access-based consumption' in their study of Zipcar. But this term is also too broad, since it encompasses both collaborative consumption and sharing. However, they make clear that they are studying collaborative consumption in noting that consumers who use the service want a special type of access: 'Instead of buying and owning things, consumers want access to goods and prefer to pay for the experience of temporarily accessing them' (Bardhi and Eckhardt 2012, p. 881). They also offer the more accurate label of 'market-mediated access'.

The push to collaboratively consume automobiles

Just as young drivers are forgoing or delaying obtaining a driver's licence, there is also an increase in the number of Americans leasing rather than buying automobiles. The percentage of leasing new cars rose from slightly over 10 per cent in 2009 to more than 27 per cent in 2013 (Rudarakanchana 2013). In Sweden, many people lease their cars through their place of employment with the fee being deducted from their salaries. Aside from not having many of the responsibilities of ownership, the lack of equity generally means that for the same payment that would be made on a car loan in the case of purchase, the person leasing a car can acquire the use of a more expensive car. Zipcar and other collaborative consumption ventures take this leveraged lifestyle trend a step further in leasing cars by the hour among those who pay a yearly membership fee. Members reserve cars online and can unlock and operate them with a smart phone. There is a credit card in the car if fuel is needed and all insurance and maintenance is handled by Zipcar. The vehicle must currently be returned to the place where it was picked up, although this may change as more locations are added, also potentially eliminating parking fees since Zipcar rents designated spaces in various parking lots and garages. In the future, with autonomous self-driving cars, perhaps the cars will come to the drivers rather than the reverse.

Noting the trend towards short-term rental, Avis recently acquired Zipcar. Besides firms like Avis, automobile manufacturers are also responding to the increasing preference for consumers to rent rather than buy vehicles. Daimler Benz (Mercedes) has started Car2Go, while BMW developed DriveNow, Volkswagen created Quicar, and Peugeot has fielded Mu, all short-term rental plans for these firms' cars (Firnkorn and Müller 2012; Steinberg and Vlasic 2013; Wüst 2011). Although this can be seen as offering an alternative to purchasing their cars, they are responding to the decreasing interest of young people in owning a car as an important part of their self-definition. Changing North American residential patterns are another factor, as young people and immigrants increasingly forsake the suburbs for the city and find that with public transportation, a car is unnecessary. The auto companies see short-term rental as a way to still be involved in serving their transportation needs (Nelson 2013; Rosenthal 2013; Wohlsen 2013). Short-term rental programmes in cities offer them a way to continue to earn money by renting out rather than selling their cars outright to individuals.

Another strategy is to appeal to existing and potential car owners by offering plans to help them rent out their cars on a short-term basis themselves to recover some of the cost of owning and operating them. General Motors offers a P2P application called Relay Rides that brings together car owners with those who wish to use the cars on a short-term basis. They also provide a blanket insurance policy that covers the car while in use. Thus, like the European car companies that have started short-term rental of their cars, GM has found a way to keep in the game, with the help of their OnStar system of in-car communication.

Besides its Car2Go rental service, Mercedes also offers Car2Gether that connects car owners and those seeking a ride to a certain location. It allows drivers to publish their planned trips online together with the fee, say, to include another passenger on a trip between Berlin and Dusseldorf. The service has not only resulted in favourable press, it also offers a car-reducing, congestion-reducing, and environmentally friendly way to make car ownership more attractive. Many such services to facilitate ride sharing have also emerged independently through companies like Uber, Local Motion, Zimride, Spride, Getaround, Lyft, Sidecar, and blablacar. They too offer ways to reduce traffic, reduce pollution, save money, and create efficiencies in comparison to cities and highways full of single-occupant vehicles.

Some ride-sharing sites have found additional incentives to encourage carpooling. San Francisco's 'casual car pool' started by setting up locations throughout the San Francisco Bay area near public transportation terminals. Those seeking rides wait at these locations and drivers stop to offer rides to different locations. Drivers who pick up a passenger can subsequently use the high occupancy vehicle lane, assuring that they get there quicker and the rider gets there quickly and inexpensively (gas money is optional). The city also uses an online reputation system to warn of bad drivers and bad passengers. Truly cooperative car-sharing organizations also exist, like Majorna in Göteborg, Sweden. In this case, the members rotate in maintaining, programming, and scheduling the use of their small car fleet. Still some of the members worry that with 300 members and 29 cars, it may be getting too big to personally know everyone in the member community (Jonsson 2007).

The role that schemes like this play in building a sense of community cannot be overstated. As Putnam (2000) has demonstrated, we seldom know our neighbours any longer. This builds a general distrust. Arguably, fears of 'stranger danger' have largely driven hitchhiking out of existence, at least until recently. Community sharing organizations, including ride sharing, together with reputation systems when the organization grows beyond the neighbourhood level, help to create safe and effective community sharing. And, as the name blablacar (a ride-sharing organization in Europe) implies, part of the pleasure of giving and receiving a ride is the conversation that very likely ensues (before meeting, drivers and passengers online can match the level of bla-bla they desire). Such systems, as well as even more distant relationships of trust like those between eBay buyers and sellers, help to restore our faith in other people and make us feel like we are not alone in a big anonymous place (Masum and Tovey 2011; Sacks 2011; Solove 2007). Some companies like TrustCloud offer a reputational system that can be used across different collaborative consumption platforms (Holson 2012).

More collaborative consumption

The success of collaborative consumption models may not provide all of the warm fuzzy feelings of true sharing, but it has benefits for individuals, communities, and the environment. There are many more types of consumption organization covering a variety of product and service needs, ranging from P2P lending, crowd funding, shared Wi-Fi, community supported agriculture, skill barter banks, car repair, and child care, to the use of designer dresses and handbags, catering, packing boxes for moving, and garden space (see Botsman and Rogers 2010; Leadbeater 2009; Slee 2013). All use the Internet to facilitate locating things for hire that we once had to buy, rent, or lease for extended periods (Cheshire *et al.* 2010; Durgee and O'Connor 1995). A Toronto car-sharing organization, AutoShare, inverts the former pro-ownership question of 'Why rent when you can buy?' by asking, 'Why own when you can rent by the hour?' The cumulative effect of this increase in non-ownership models may just be to begin to shift us from the ownership paradigm into something a little less possessive, a little more fluid, and a lot more socially responsible, even if we do it for the sake of saving money, having a nicer handbag or car than we could afford to buy, and meeting new people to talk to and perhaps even develop a relationship with.

This doesn't mean that collaborative consumption and sharing are without problems. The music and film industries vigorously opposed online sharing of their products. The movie, music, and publishing industries have been slow to embrace non-ownership models and often go to elaborate legal and technological lengths to discourage copying and sharing of their content. Hotels are pressuring cities, states, and provinces to regulate room letting through services like Airbnb, House-Trip, Windu, and 9flats in the same way they regulate hotels and old-fashioned bed and breakfast places. Restaurants in cities like Hong Kong are doing the same to get 'private kitchen' paid-for meals in people's homes regulated in the same way that other commercial food service locations are inspected and licensed. Likewise, some banks and savings and loan organizations seek regulations of P2P lending and crowd funding alternatives. But such reactionary tactics were not effective in the music and film industries and will probably not be effective in these contexts either. The spirit of sharing and collaborative consumption that has been stirred ever since Napster is just too strong. The sharing that takes place in social media, blogs, Wikis, web pages, and forums lets us rightly feel that there is a big world of 'free stuff' out there on the Internet and that perhaps we should do our bit to contribute to it as well. Given the rapid growth of many of these Internet-facilitated collaborative consumption and sharing enterprises, and their benefits to the environment as well as all concerned, with the exception of the industries that they threaten, it may be time for traditional businesses to rethink their business models. The next section explores how this might be done and the incentives for considering business models based non-ownership alternatives.

Business implications

The initial reaction of many businesses to sharing and collaborative consumption trends is that they should be resisted because they lead to a decline in purchases as well as emphasize shared ownership or short-term rental (Boesler 2013). The histories of the music, film, and publishing industries and their largely futile fight to use IPR legislation to stop such ventures, offer good case studies of how not to react to a changing economy. Rather than fight new ventures and alienate consumers in the process, it is time to think about how to embrace new technologies and opportunities based on collaborative consumption and sharing.

The short-term car rental and ride-sharing efforts by a number of the automobile and car rental companies offer a good alternative example of the benefits of embracing non-ownership business models. Even in the music and film industries, some, largely new, companies offer paid downloads, streaming, and video on demand. Netflix, Internet service providers, cable television companies, Apple, Amazon, and most software companies have all become involved in selling digital access rights rather than ownership of tangible products. Some companies like Amazon offer both, providing consumers with a choice. Greeting cards, mail, newspapers, magazines, photography services, and many other traditional ways of doing business have already been substantially changed by new intangible as well as sharing and collaborative consumption offerings. Some, like Google and Baidu (in China), have managed to adapt and survive or flourish online, for example by offering free content and gaining revenue from banner advertising on their sites. Others have not and many of those who have been slow to change have not survived. Those traditional companies that are still in business are attempting to hang on to a shrinking segment of the market willing to stick to old ways and to pay full price for what other consumers can get free or for far less than the purchase price. Those consumers who have high ethical qualms, those who are cash rich and time poor, and those who lack technical expertise may prefer traditional ownership and paid downloads, but these are not the consumers of the future. They tend to be older consumers who are migrants to the digital world as opposed to younger consumers who were 'born digital'.

An alternative business model for some successful, again largely new, firms is to provide free content and look for other sources of revenue. Google, Twitter, Facebook, and other Internet service companies don't charge for their services and encourage people and companies to use them at no cost. Their substantial revenues instead come from advertising by firms who relish the ability to target messages to users who are matched in terms of expressed interests in their offerings. Publishers and the music industry also increasingly rely on libraries to buy subscriptions and e-books and then provide them free to patrons. Music distribution services (again mostly new companies) like iTunes, Rhapsody, and Spotify charge for access to their products or services rather than tangible sales of products. They in turn rebate royalties to artists and music companies. Libraries rebate copy fees to copyright holders in a similar fashion.

If a company is unable to quickly respond to the changing, dematerializing, sharing, and collaborative markets, they may still be able to compete by buying up new ventures as Avis did with Zipcar. Even digital players like Apple, Microsoft, Google, and Amazon are continuously buying up smaller specialized firms in order to improve their products and services and stay competitive. For Internet start-ups, the strategy is instead to look for new models of digitally aided distribution. Examples here are Bag Borrow or Steal and Rent the Runway, which offer rotating access to designer handbags, designer dresses, and accessories for an affordable monthly fee. The number of consumers who can afford to purchase such goods is relatively small, but the number of those who are attracted to such bling when it can be rented for an affordable fee is potentially much larger. And, unlike those who buy counterfeit goods, they are assured of getting the real thing. Such businesses thus expand the market to include those who long for goods they can ill afford. Even in pre-digital days, this was the business concept behind time-share condominiums and automobile leasing.

Before getting too euphoric about the new sharing and collaborative consumption markets, the business opportunities they may provide, and the favourable impact on sustainable consumption they may facilitate, it is also wise to step back and consider whether these are truly promising and long-term trends. Although the parallel rise of reuse and recycling seem promising (Ekström and Salomonson 2014), Bloemer (2001) found that the used clothing given to large charitable organizations and resold to wholesalers ends up in less-affluent countries like Zambia where the cheap prices for this used clothing ended up destroying the local textile industry. We need to look beyond the surface of feel-good effects of increased sharing as well. For example, we found that among school children, the ethos of sharing also extends to sharing exam answers (Belk and Llamas 2012). So sharing is not necessarily unequivocally good.

We might ask, will these new business and consumption models continue to flourish as world economies improve? Will 'Generation Y' or 'Millenials' continue to prefer living downtown and not owning cars or the other possessions that fill suburban homes as they get partners and have children? Or will currently low interest rates and better highways foster more suburban home buying as economies recover? Or are sharing and collaborative consumption, like the digital revolution, genies that cannot be put back into the bottle? Those who live in rural areas have generally had to rely on each other and have learned to trust and share more than those in big anonymous cities. But the Internet, reputation economies, and new forms of sharing and collaborative consumption seem to have brought new ways and new levels of trust of strangers online. So long as safeguards continue to be developed to thwart online outright theft and scams, the future looks very promising. We have not yet given up our love of ownership, as surging demand for rented storage space in North America, Europe, and Asia demonstrate. However, it does seem that the bloom is off the rose of ownership and alternative forms of access are growing.

At a more general level, we should consider the impact of trends such as global warming, rising fuel and raw material prices, growing populations, rising pollution levels, and other anticipatable trends affecting future sharing and collaborative consumption trends as well as the attractiveness of living in cities. As white collar jobs are untethered from offices and as telecommuting becomes more common, it is conceivable that city living, despite all its attractions, will begin to lose its allure. All of this must also be considered in predicting the future of sharing and collaborative consumption.

Conclusions

In North American and Western Europe we use our cars 8 per cent of the time (Sacks 2011). As noted earlier, electric drills are used for 6 to 20 minutes in their lifetime (Earth Share, no date; Steffen 2007). Sharing makes sense for the consumer, the economy, the environment, and the community. It can also make sense for businesses, as I have argued above. According to some proclamations, collaborative consumption could be as important a development as the Industrial Revolution (Botsman and Rogers 2010). Along with the digital revolution, sharing and collaborative consumption have already begun to change the way we think about ownership.

Some of our best brightest thinkers are creating new Internet start-ups based on offering disruptive technologies that undermine businesses steeped in models of selling products to consumers. For both consumers and businesses, it is increasingly imperative to think about sharing and collaborative consumption as alternative ways of consuming and doing business. They offer hope for all of us that there are easy and practical ways to move towards more sustainable consumption.

References

Aigrain, Philip (2012), *Sharing: Culture and the economy in the internet age*, Amsterdam: Amsterdam University Press.
Arnold, Jeanne E. and Ursula A. Lang (2007), 'Changing American home life: Trends in domestic leisure and storage among middle-class families,' *Journal of Family Economic Issues*, 28: 23–48.

Arsel, Zeynep and Susan Dobsha (2011), 'Hybrid pro-social exchange systems: The case of Freecycle,' *Advances in Consumer Research*, 39.

Bardhi, Fleura and Giana Eckhardt (2012), 'Access based consumption: The case of car sharing, *Journal of Consumer Research*, 39, 881–898.

Belk, Russell (1988), 'Possessions and the extended self', *Journal of Consumer Research*, 15, 139–168.

Belk, Russell (2007), 'Why not share rather than own?' *Annals of the American Academy of Political and Social Science*, 611, 126–140.

Belk, Russell (2010), 'Sharing', *Journal of Consumer Research*, 36, 715–734.

Belk, Russell (2013), 'Extended self in a digital world', *Journal of Consumer Research*, 40, 477–500.

Belk, Russell (forthcoming), 'Sharing versus pseudo-sharing in Web 2.0', *The Anthropologist*, 4(2).

Belk, Russell and Rosa Llamas, (2012), 'The nature and effects of sharing in consumer behavior,' in David Mick, Simone Pettigrew, Connie Pechmann, and Julie Ozanne (eds), *Transformative Consumer Research for Personal and Collective Well-being*, New York: Routledge, 625–646.

Benkler, Y. (2004), 'Sharing nicely: On shareable goods and the emergence of sharing as a modality of economic production', *Yale Law Journal*, 114, 273–358.

Bloemer, Sandra (2001), *T-Shirt Travels: A documentary on secondhand clothes and third world debt in Zambia*, Boston: PBS Films.

Boesler, Mathew (2013), 'The rise of the renting and sharing economy could have catastrophic ripple effects', *Business Insider*, August 12. Available online at http://www.businessinsider.com/rise-of-the-renting-and-sharing-economy-2013-8?op=1 (accessed 18 August 2014).

Botsman, Rachel and Roo Rogers (2010), *What's Mine Is Yours: The rise of collaborative consumption*, New York: Harper Collins.

Brooks, Bianca (2013), 'Teens use Twitter to thumb rides', *all tech considered*, August 15. Available online at http://www.npr.org/blogs/alltechconsidered/2013/08/15/209530590/teens-use-twitter-to-thumb-rides (accessed 18 August 2014).

CBS (2009), 'Poll: Young say file sharing OK', *CBS News*, February 11. Available online at http://www.cbsnews.com/stories/2003/09/18/opinion/polls/main573990.shtml (accessed 18 August 2014).

Cheshire, Lynda., Peter Walters, and Ted Rosenblatt (2010), 'The politics of housing consumption: Renters as flawed consumers on a master planned estate', *Urban Studies*, 47, 2597–2614.

Durgee, Jeffrey and Gail O'Connor (1995), 'An exploration into renting as consumption behavior', *Psychology and Marketing*, 12, 89–104.

Earth Share (no date), 'A new economy based on sharing'. Available online at http://www.earthshare.org/2012/05/sharing.html (accessed 18 August 2014).

Ekström, Karin and Niklas Salomonson (2014), 'Reuse and recycling of clothing and textiles— A network approach,' *Journal of Macromarketing*, 34(3), 383–399.

Firnkorn, Jörg and Martin Müller (2012), 'Selling mobility instead of cars: New business strategies of automakers and the impact on private vehicle holding,' *Business Strategy and the Environment*', 21, 264–280.

Fitzsimmons, James (1985), 'Consumer participation and productivity in service operations', *Interfaces*, 15, 60–67.

Galbreth, Michael, Bikram Ghosh, and Mikhael Shor (2012), 'Social sharing of information goods: Implications for pricing and profits,' *Marketing Science*, 41 (July–August), 603–620.

Gansky, Lisa (2010), *The Mesh: Why the future of business is sharing*, New York: Portfolio Penguin.

Giesler, Markus (2006), 'Consumer gift systems', *Journal of Consumer Research*, 32, 283–290.

Giesler, Markus (2008), 'Conflict and compromise: Drama in marketplace evolution', *Journal of Consumer Research*, 34, 739–753.

Grassmuck, Volker (2012), 'The sharing turn: Why we are generally nice and have a good chance to cooperate our way out of the mess we have gotten ourselves into', in Wolfgang Sützl, Felix Stalder, Ronald Maier, and Theo Hug (eds), *Cultures and Ethics of Sharing*, Innsbruck: Innsbruck University Press, 17–34.

Hemetsberger, Andrea (2012). ''Let the source be with you!'—Practices of sharing in free and open-source communities', in Wolfgang Sützl, Felix Stalder, Ronald Maier, and Theo Hug (eds), *Cultures and Ethics of Sharing*, Innsbruck: Innsbruck University Press, 117–128.

Hennig-Thurau, Thorstein, Viktor Henning, and Henrik Sattler (2007), 'Consumer file sharing of motion pictures', *Journal of Marketing*, 71, 1–18.

Holson, Laura (2012), ''What were you thinking?' For couples, new source of online friction', *New York Times*, April 25, online edition.

Humphreys, Ashley and Kent Grayson (2008), 'The intersecting roles of consumer and producer: A critical perspective on co-production, co-creation and prosumption', *Sociological Compass*, 2, 963–980.

Ingold, Tim (1986), *The Appropriation of Nature: Essays on human ecology and social relations*, Manchester: Manchester University Press.

John, Nicholas (forthcoming), 'The social logics of sharing', *The Communication Review*.

Jonsson, Pernilla (2007), 'A tale of a car sharing organization (CSO) monster', in Helen Brembeck, Karin Ekström, and Magnus Mörck (eds), *Little Monsters: (De)coupling assemblages of consumption*, Berlin: Lit Verlag, 149–164.

Lamberton, Cait and Randall Rose (2012), 'When is ours better than mine? A framework for understanding and altering participation in commercial sharing systems', *Journal of Marketing*, 76, 109–125.

Lanier, Clinton, Jr. and Hope Schau (2007), 'Culture and co-creation: Exploring consumers' inspirations and aspirations for writing and posting on-line fan fiction', in Russell Belk and John Sherry, Jr. (eds), *Consumer Culture Theory: Research in consumer behavior*, Vol. 11, Amsterdam: Elsevier, 321–342.

Leadbeater, Charles (2009), *We-Think: Mass innovation, not mass production*, London: Profile Books.

Lessig Lawrence (2008), *Remix: Making art and commerce thrive in the hybrid economy*, New York: Penguin.

Masum, Hassan and Mark Tovey (eds) (2011), *The Reputation Society: How online opinions are reshaping the offline world*, Cambridge, MA: MIT Press.

McCartney, Kelly (2012), 'New notes on solidarity among sharers', June 13. Available online at http://www.shareable.net/blog/new-study-notes-solidarity-among-sharers (accessed 18 August 2014).

Mont, O. K. (2002), 'Clarifying the concept of product service system', *Journal of Cleaner Production*, 10, 237–245.

Nelson, Noah (2013), 'Why millennials are ditching cars and redefining ownership', National Public Radio Morning Edition, August 21. Available online at http://www.npr.org/2013/08/21/209579037/why-millennials-are-ditching-cars-and-redefining-ownership (accessed 18 August 2014).

Ozanne, Lucie and Paul Ballantine (2010), 'Sharing as a form of anti-consumption? An examination of toy library users', *Journal of Consumer Behaviour*, 9, 485–498.

Polyani, Karl (1944), *The Great Transformation*, Boston: Beacon Press.

Postigo, Hector (2003), 'Emerging sources of labor on the Internet: The case of America online volunteers', *International Review of Social History*, 48 (December), 205–223.

Prahalad, C. K. and Venkat Ramaswamy (2004), 'Co-creation experiences the net practice in value creation', *Journal of Interactive Marketing*, 18, 5–14.

Putnam, Robert (2000), *Bowling Alone: The collapse and revival of American community*, New York: Simon and Schuster.

Reagle, Michael, Jr. and Lawrence Lessig (2010), *Good Faith Collaboration: The culture of Wikipedia*, Cambridge, MA: MIT Press.

Ritzer, George and Nathan Jurgenson (2010), 'Production, consumption, prosumption: The nature of capitalism in the age of the digital "prosumer"', *Journal of Consumer Culture*, 10, 13–36.

Rosenthal, Elizabeth (2013), 'The end of car culture', *New York Times*, June 29, online edition. Available online at http://www.nytimes.com/2013/06/30/sunday-review/the-end-of-car-culture.html?pagewanted=all&_r=0 (accessed 18 August 2014).

Rudarakanchana, Nat (2013), 'More people than ever get cars and auto vehicles by leasing, according to Experian report', *International Business Times*, June 6, online edition. Available online at http://www.ibtimes.com/more-people-ever-get-cars-auto-vehicles-leasing-according-experian-report-1293279 (accessed 18 August 2014).

Sacks, Danielle (2011), 'The sharing economy', *Fast Company*, April 18, online edition. Available online at http://www.fastcompany.com/1747551/sharing-economy (accessed 18 August 2014).

Slater, Don (2000), 'Consumption without scarcity: Exchange and normativity in an Internet setting', in Peter Jackson, Michelle Lowe, Daniel Miller, and Fran Mort (eds), *Commercial Cultures: Economies, practices, spaces*, Oxford: Berg, 123–142.

Slee, Tom (2013), 'Why the sharing economy isn't', *Whimsley* (blog), August 30. Available online at http://tomslee.net/2013/08/why-the-sharing-economy-isnt.html (accessed 18 August 2014).

Solove, Daniel (2007), *The Future of Reputation: Gossip, rumor, and privacy on the Internet*, New Haven, CT: Yale University Press.

Steffen, Alex (2007), 'Use community: Smaller footprints, cooler stuff and more cash', *World Changing*, February 15. Available online at http://www.worldchanging.com/archives/006082.html (accessed 18 August 2014).

Steinberg, Stephanie and Bill Vlasic (2013), 'Car-sharing services grow, and expand options', *New York Times*, January 25, online edition. Available online at http://www.nytimes.com/2013/01/26/business/car-sharing-services-grow-and-expand-options.html (accessed 18 August 2014).

The Economist (2013), 'The rise of the sharing economy: On the Internet everything is for hire', March 9. Available online at http://www.economist.com/news/leaders/21573104-internet-everything-hire-rise-sharing-economy (accessed 18 August 2014).

Toffler, Alvin (1980), *The Third Wave*, New York: William Morrow.

Willer, Robb, Francis Flynn, and Sonya Zak (2012), 'Structure, identity, and solidarity: A comparative field study of generalized and direct exchange', *Administrative Science Quarterly*, 57, 119–155.

Wittel, Andreas (2011), 'Qualities of sharing and their transformation in the digital age', *International Review of Information Ethics*, 15. Available online at http://www.i-r-i-e.net/inhalt/015/015-Wittel.pdf (accessed 18 August 2014).

Wohlsen, Marcus (2013), 'What the sharing economy needs is a little less democracy', August 28, *Wired*, online edition. Available online at http://www.wired.com/2013/08/sharing-economy-localmotion/ (accessed 18 August 2014).

Wüst, Christian (2011), 'German automakers embrace auto-sharing', *Der Spiegel*, November 10, online English edition. Available online at http://www.spiegel.de/international/business/driving-force-german-automakers-embrace-car-sharing-a-796832.html (accessed 18 August 2014).

13
SUPPLEMENTING THE CONVENTIONAL 3R WASTE HIERARCHY

Considering the role of carbon rationing

Maurie J. Cohen

Introduction

The specter of rationing is a bitter pill for most consumers and immediately evokes images of scarcity and deprivation. Even hypothetical fiats to regulate consumption typically trigger mental experiments to devise imaginative procedures to circumvent prospective measures. Probably in no small measure, fears of extremely negative public reactions have contributed to the reluctance of policy makers in most countries to even discuss consumption controls as a means to reduce greenhouse-gas emissions or to address other contemporary environmental problems (Cohen 2011; Cox 2013) Rationing is the policy tool about which we dare not speak. Interventions have instead focused on improving technical efficiencies (automotive fuel-economy standards), implementing economic incentives to more closely align the behavior of producers and consumers with biophysical limits (so-called cap and trade programs), and trialing ameliorative technologies (underground sequestration of carbon) (Huesemann and Huesemann 2011; Lipschutz 2012; see also Morozov 2013).

Release by the Intergovernmental Panel on Climate Change (IPCC) of its Fifth Periodic Assessment in September 2013 has enlivened a dormant debate on the efficacy of consumption controls as a decarbonization strategy. In this document, the IPCC announced that to keep the expected increase in average global temperatures from reaching an unacceptably dangerous level it will be necessary to prevent cumulative carbon-dioxide releases from exceeding a threshold volume. With the passage of time, we may come to realize that this pronouncement was the first step toward supplementing the customary three-point environmental management maxim of "reduce, reuse, recycle" with a fourth alliterative stratagem—"ration."

Any meaningful consideration of rationing needs to begin by differentiating between *price* and *non-price* consumption controls. In the first instance, prices are the primary means for allocating scarce resources in a market economy. To take an extreme example, the high price of a Lamborghini allocates available supplies to a relatively small number of buyers with the financial means to purchase such a car. But price rationing is not only used to manage the distribution of expensive Italian sports cars; it is also generally regarded as an effective way to apportion all manner of products ranging from apples to iPods. This is not to say that the resultant prices lead to socially desirable results. In cases where supplies are constrained or there is a long lead time to augment available inventories (housing being the archetypal example), wealthier individuals can bid up prices beyond the financial reach of less affluent counterparts. In fact, one of the tasks of political institutions in market economies is to intervene in various ways to secure outcomes that are deemed to be more optimal than would otherwise occur on the basis of price rationing alone.

By contrast, non-price interventions constitute the mode of consumption control that most people understand as "rationing." Rather than allow prices to serve as the means for determining allocation, non-price rationing seeks to regulate distribution by managing quantities. While the far-reaching application of non-price rationing is generally reserved for emergencies brought about by war or severe supply disruptions due, for instance, to labor strikes or natural disasters, it is worth noting that consumers encounter modest applications of this policy tool on an everyday basis. Minimum age requirements on the purchase of alcohol represent a non-price strategy for discouraging youth drinking. Prescription drugs, firearms, and cigarettes are other consumer products that most countries subject to consumption controls of different degrees of stringency. The point here is that non-price rationing is not as exceptional as we are sometimes led to believe and it is often invoked without exceptional controversy.

This chapter seeks to assess the status of consumption rationing in contemporary environmental and sustainability policy debates. The discussion unfolds as follows. The next section reviews the current political status of rationing as a means to regulate greenhouse-gas emissions. The third section explains how the recent decision by the IPCC is likely to rekindle consideration of consumption controls as a way to reduce carbon releases in affluent nations. The fourth section examines the willingness of consumers to "buy" into rationing and determines that the obstacles to its implementation may not be as formidable as we might initially surmise. The conclusion considers some developments related to an emergent "post-consumerist" shift that might lessen public opposition to consumption controls as a climate-management strategy.

Assessing the current status of personal carbon rationing

The UK is prominent as the country that to date has most intensively considered rationing as a potential policy intervention to reduce its greenhouse-gas emissions (Fawcett et al. 2007; Fawcett 2012; House of Commons 2008; DEFRA 2008; Cohen 2011). This interest has principally taken the form of personal carbon allowances, a concept publicly introduced in a 2006 lecture by the Environment

Secretary at the time, David Miliband. Brother of the present leader of the British Labour Party and erstwhile competitor for the position himself, Miliband (2006) used this opportunity to encourage people to

> Imagine a country where carbon becomes a new currency. We carry bank cards that store both pounds and carbon points. When we buy electricity, gas, and fuel, we use our carbon points, as well as pounds. To help reduce carbon emissions, the Government would set limits on the amount of carbon that could be used. Imagine your neighbourhood. Each neighbour receives the same free entitlement to a certain number of carbon points. The family next door has an SUV and realize they are going to have to buy more carbon points. So instead they decide to trade in the SUV for a hybrid car. They save 2.2 tonnes of carbon each year.

During the aftermath of this presentation, the UK's Department of Environment, Food and Rural Affairs (DEFRA) began a research program centered specifically on carbon rationing that over the next few years explored both the social and technological dimensions of how such a system might work. For instance, studies were commissioned on the design of small data-storage devices that people would use to manage their individual carbon accounts as well as the practicability of various allowance schemes. In the wake of a modest demonstration project, media reports for a time even heralded the distribution of specially issued carbon-denominated "smart" cards on a countrywide basis by 2013 (Adam 2007). The focus evolved during this period—both terminologically and operationally—from "personal carbon rationing" to "personal carbon allowances" and "personal carbon trading" to overcome connotations of scarcity and to suggest instead both entitlement and conformance with prevailing market-oriented prerogatives. There exists today a quite sizeable body of academic literature on the subject, much of it spawned by funding streams that emanated from efforts launched in connection with this specific upwelling of interest.[1]

This attention by the UK government catalyzed numerous grassroots experiments across the country in household- and community-based carbon management and some of these ideas diffused more widely. A notable expression took the form of carbon rationing action groups (CRAGs) comprising small groups of people that encouraged participants to initiate and maintain low-carbon lifestyles (Paterson and Stripple 2010; Howell 2012). The "transition town" movement is another related development whereby activist residents formulate plans that typically entail the imposition of voluntary consumption controls (Hopkins 2008; Seyfang and Haxeltine 2012). For instance, participants might collectively agree to reduce their weekly reliance on a personal automobile, or forsake its use altogether. At the level of international environmental politics, the concept of "contraction and convergence" garnered attention during this time as a strategy for drawing down greenhouse-gas emissions in affluent nations and proportionately raising releases in poorer nations to foster greater global equity (Meyer 2000; Stott 2012).

Since 2008, political developments in the UK and the international financial crisis have diminished the visibility of carbon rationing. Miliband left his post as Environment Secretary and during the last three years of the Labour government served as Secretary of State for Foreign and Commonwealth Affairs where his portfolio limited the amount of attention he was able to devote to the issue. Following his departure, DEFRA issued a "pre-feasibility analysis" that expressed ongoing interest in carbon rationing, but ultimately concluded that it was an idea that was "ahead of its time." Formation of the Conservative–Liberal Democrat coalition government in 2010 then moved Miliband into the opposition and forced him to compete unsuccessfully against his brother for leadership of the Labour Party. David Miliband is not presently involved in British politics.

Renewing interest in carbon rationing

A rationing regime is premised on two basic design principles: a limit on aggregate consumption and a formula for apportioning available quotas. However, a workable program requires consideration of a variety of other more nuanced factors including the provision of supplementary allocations to certain subpopulations (e.g. pregnant women and heavy laborers), the creation of mechanisms to trade or exchange rations, and the implementation of dispensations to allow for home or community provisioning. These, though, are relatively incidental matters that can be set aside for the time being.[2]

The IPCC's Fifth Periodic Assessment delivered the sobering news that a cap on carbon emissions of one trillion metric tons would be necessary to contain the increase in average global temperature to 2°C. The assertion is that beyond this threshold climate change will become more severe and unstable and exact increasingly onerous economic and ecological costs. The historic production of greenhouse gases (since the start of the Industrial Revolution) already accounts for approximately half of this budget and globally we are on track to exceed this newly declared ceiling by 2030. The first piece of a carbon-rationing regime is thus, at least in scientific terms, established and represents an authoritatively imposed biophysical limit on fossil-fuel-driven activities. The next step will be to develop a plan for distributing the remaining 500 billion[3] metric tons of carbon emissions. This is likely to be a protracted and conflictive political process with relatively wealthy (and soon-to-be wealthy) consumers stridently wrangling to retain "sovereignty" over their emission streams. Several specific challenges will need to be overcome to accomplish this second step of apportioning individual carbon shares.

First, as Stan Cox (2013) observes, it is necessary to differentiate between "defensive" and "offensive" rationing (see also Bentley 1998). In short, defensive rationing protects dwindling supplies when opportunities to restock are limited or uncertain, while offensive rationing is motivated by a proactive desire to avoid future adversity. Experience with consumption controls over the past century suggests that defensive measures that are strategically and fairly applied in the face of clear and unambiguous perils can be quite successful.[4] While acknowledging regulatory lapses and black-market trading, scholars of British rationing during and after World War II offer generally positive evaluations of its overall effectiveness (see, e.g., Zweiniger-Bargielowska 2000; Longmate 2002; Connelly 2004).[5] However, similar consumer regulation as a response to climate change would need—at least for the next couple of decades—to be implemented on an offensive basis. In other words, consumption controls would have to be adopted and maintained as precautionary rather than protective measures and this means that ensuring compliance would likely require extremely vigorous—as well as creative—enforcement.

Second, to reduce greenhouse-gas emissions by an appreciable amount it would be necessary for consumer regulations on carbon to remain in place for an extended (and probably indefinite) period. This situation stands in stark contrast to prior experience where demand regulations have typically been imposed as temporary measures. The UK, where consumption controls were variously in place from 1939 until 1955, represents an extreme example from the annals of modern history. This length of time, though, does not begin to approach what would be required as a response to climate change. While the carbon-rationing regime might initially be launched as an emergency intervention, it would need to become normalized as an enduring condition.

Third, previous uses of consumption controls have targeted explicit categories of goods—food, fuel, clothing, furniture, and so forth. A system to limit carbon emissions would be similarly straightforward to the extent that it was confined to direct fossil-fuel purchases (e.g. gasoline at the pump), but would become much more complicated as the carbon embodied in products came to be incorporated (as surely it would need to) into consumer regulations. The methods used to calculate these indirect sources of greenhouse-gas emissions are contestable and, at least at present, largely incomprehensible to ordinary consumers.[6]

Finally, building up enforcement capacity will require sustained political commitment that will need to transcend customary partisan divisions. It would be extremely difficult to reinvigorate a carbon-rationing regime after a lapse, so resolve would need to be bureaucratically institutionalized to withstand the ebbs and flows of electoral politics. It would be necessary for public commitment for consumption controls on carbon to approach a level of compliance that prevails today with respect to, say, guns in Japan or the UK. In the absence of a sharp cultural turn, it would be difficult to achieve this target and such circumstances raise deep questions about the capacity of nominally democratic polities to maintain durable limits on carbon in the face of a serious, but nonetheless prospective and ultimately indeterminate, threat.

But will consumers buy it?

What might daily life look like for people in affluent nations under consumption controls designed to transition to a decarbonized future? An effort to glimpse a view requires clarification about the strictness of the regulatory system which, to have any marked effect, would need to be at least moderately restrictive and subject to continuous augmentation over time. Let us briefly consider three domains: energy, mobility, and food. I also offer a few cursory observations about macroeconomic changes that would likely ensue with respect to household provisioning.

First, carbon rationing would have roughly similar impacts as a steep tax on fossil fuels. The policy objective would be to undermine the economic viability of coal, oil, and natural gas as primary sources of energy and to supplant them with renewable alternatives. In particular, solar and wind power would be installed through a combination of utility-scale provisioning and micro and distributed generation. Cities and their surrounding regions, as well as the building stock, would gradually be retrofit to accommodate lifestyles organized around lower energy throughput dictated by revised design requirements and new standards of comfort and convenience (Chappells and Shove 2005; Gomi et al. 2007; Keirstead 2008; Alderson et al. 2012; Eames et al. 2013).

Second, personal automobile use would be curtailed in response to rising operational costs, decreasing fuel availability, and patterns of increasing urbanization. Daily activities would become more geographically circumscribed as accessibility took on new significance and intermodality became a more prominent feature of transportation planning. The reduction in car dependency would likely generate important public health co-benefits as non-motorized modes—notably walking and cycling—became more realistic and widely adopted alternatives. Novel forms of recreation, and new forms of leisure travel in particular, would develop in response to the increasing cost of commercial aviation (Larsen and Höjer 2007; Sgouridis et al. 2011; Grahn et al. 2013). Virtual reality tourism might come to displace the need for physical transport to distant locations (Zhang and Zhu 2012; Williams 2013).

Third, carbon rationing would encourage (re)localization of food production and consumption as globalized supply chains gave way to proximate cultivation (Powell and Lenton 2012; Laestadius et al. 2013; Kirveennummi et al. 2013; Lombardini and Lankowski 2013). Carbon-intensive foods, most notably meat, would cease to be a staple of everyday diets (Vinnari and Tapio 2009). It is also the case with respect to food that important health improving co-benefits would develop as fresh produce replaced highly processed products (Scarborough et al. 2012; Yip et al. 2013; Lang and Barling 2013).[7]

Finally, consumption controls would drive changes in the macroeconomy as livelihoods became less reliant on physical goods and dematerialized services became important sources of value creation and employment generation (Schandl and Turner 2009; Fujimori and Masui 2011; Waddock and McIntosh 2011; Frye-Levine 2012). Such circumstances could also conceivably lead to increasing diffusion of practices predicated on shared use of durable goods as an alternative to private ownership, at least for certain product categories that can be cost-effectively transacted interpersonally or via the Internet (Belk 2007, 2014; see also Belk, this volume). "Post-consumerist" social innovations encouraging reskilling, collaborative consumption, prosumption, and self- and communal provisioning could also scale up and radiate outward to a point where they became societally consequential (Crawford 2010; Botsman and Rogers 2010; Schor 2011; Cohen 2013). Indeed, there are communities in New York City and San Francisco where certain agglomeration effects have begun to take hold and to create nodal concentrations of these activities. Considerable interest in recent years has focused specifically on the so-called "maker movement" which brings together a number of these trends and is heavily centered on exploiting the potential of increasingly affordable 3-D printing applications (Anderson 2012; Hatch 2013; von Busch 2013; Gobble and Euchner 2013). The hopeful vision is that this technology will enable households to become custom manufacturers of bespoke products. Explorations on how these novel routines might converge and develop is explored in recent scenario exercises to identify sustainable lifestyle pathways (see, e.g., Mont et al. 2014; Neuvonen et al. 2014).

Conclusion

A post-consumerist shift has the potential to confer certain advantages, especially in terms of enabling people to embed their lifestyles in more diversified portfolios of economic activities than is typically common today. The prevailing—and oftentimes singularly available—practice entails selling labor in exchange for cash that is then used to purchase everyday commodities. Post-consumerism holds the prospect of a complex entwining of market exchanges, home production, and community-based fabrication. While this may be an attractive vision, it remains, at least for the vast majority of people, a distant apparition. At the same time, it is difficult to envisage a critical mass enthusiastically embracing the aforementioned future lifestyle scenario given its variance with contemporary mainstream practices and aspirations. Concomitantly, to mobilize the public around carbon rationing would arguably require cultivation of a domestic political condition tantamount to total war and creation of an array of consumer regulations that were but one part of a more expansive policy package (Cohen 2011; Cox 2013). However, we should not underestimate the potential of instability and anxiety—as well as changing economic circumstances and livelihood incentives—to reorient common sensibilities over the next couple of decades (Dickinson 2009; Hanlon et al. 2011). This will especially be the case if it turns out that forecasts have underestimated the adversity induced by rising global temperatures. It is probably unrealistic at present to actively consider a policy program premised on consumption controls to transition to a low-carbon society, but preliminary preparations to do so could prove to be a useful anticipatory measure and would be justified by current levels of uncertainty. In this sense, it might be helpful to consider carbon rationing in much the same way that we treat oceanic iron fertilization and other geoengineering strategies. These are emergency interventions to be deployed only in the event that future circumstances become sufficiently critical, but in the meantime merit preparatory experimentation.

It may also be the case that the overall process of designing and implementing a rationing regime will be less politically controversial and administratively onerous if eventual interventions can amplify rather than obstruct ongoing social developments and evidence is emerging that such fortuitous conditions may be coming into view. In virtually all affluent countries, the aging of the "baby boom" generation is driving a shift in aggregate household expenditures away from material goods and toward healthcare, leisure recreation, and other less resource-intensive services. Many of these nations are experiencing a related renewal of interest in urban lifestyles and such preferences are raising questions about the future viability of the familiar suburban model. Notably, the centrality of the personal automobile is being displaced by new economic realities and cultural values. Increasing indications in the United States and several European countries suggest that car use has already begun to decline (a phenomenon being referred to as "peak car") due to a combination of societal aging, financial constraints, and novel attitudes among youth. These developments are beginning to weaken the social validation afforded to prevalent automobile use and to shape an incipient process of social problematization.[8] There is also evidence, at least in some quarters, that similar changes are overturning seemingly indomitable cultural ideals regarding outsized homes, suburban sprawl, and geographically extensive lifestyles.

If these new consumer attitudes become even moderately entrenched, a carbon-rationing regime could ultimately be regarded as a relatively benign policy tool that reinforces rather than resists ascendant practices and routines. Short of actual implementation, there is thus much that can be done to create the preconditions for the potential establishment of consumption controls to facilitate the transition to a low-carbon society.

Notes

1 Refer to the special issue (10:4) of the journal *Climate Policy* from 2010 for a useful introduction on personal carbon rationing.
2 For treatment of the administrative details on establishing a system of consumption controls see Zweiniger-Bargielowska (2000), Cox (2013), and the extensive body of work by social historians on rationing in the UK during and after World War II. Other episodes of rationing have also received consideration, most notably the gasoline crises of the 1970s in the United States and the economic reorganization that unfolded in Cuba during the early 1990s after the collapse of the Soviet Union.
3 Throughout this chapter, billion means 10^9 or a thousand million.
4 Assessments of the success of consumption controls implemented in the United States during this period come to more ambiguous conclusions that are largely attributable to the ambivalence of the American government for rationing and the uneven way in which policies were applied.
5 It is worth noting that there exists a more critical literature on the effectiveness of wartime rationing in the UK. See, in particular, Calder (1995).
6 Development of accounting methodologies to produce consumption-based greenhouse-gas measures is currently an active area of inquiry. For recent examples, see Peters (2008) and Barrett *et al.* (2013).
7 Prior applications of consumption rationing have been linked to improvements in public health. As Zweiniger-Bargielowska (2000) describes, during World War II the imposition

of strict price controls in combination with ensured availability of fixed quantities improved average British dietary standards. See also Huxley et al. (2000).
8 See, for example, a recent article by Douglas et al. (2011) that provocatively asks whether "cars are the new tobacco."

References

Adam, D. (2007), "Carbon credits: Rationing project tests government plans to make pollution personal," *The Guardian*, 10 September. Available online at http://www.theguardian.com/environment/2007/sep/10/climatechange.politics (accessed 18 August 2014).

Alderson, H., G. Cranston, and G. Hammond (2012), "Carbon and environmental footprint of low carbon UK electricity futures to 2050," *Energy*, 48(1), 96–107.

Anderson, C. (2012), *Makers: The new industrial revolution*, New York: Crown Business.

Barrett, J., G. Peters, T. Wiedmann, K. Scott, M. Lenzen, K. Roelich, and C. Le Quéré (2013), "Consumption-based GHG emission accounting: A UK case study," *Climate Policy*, 14(4), 451–470.

Belk, R. (2014), "You are what you can access: Sharing and collaborative consumption online," *Journal of Business Research*, 67(80), 1595–1600.

Belk, R. (2007), "Why not share rather than own?" *Annals of the American Academy of Political and Social Science*, 611, 126–140.

Bentley, A. (1998), *Eating for Victory: Food rationing and the politics of domesticity*, Champaign, IL: University of Illinois Press.

Botsman, R. and R. Rogers (2010), *What's Mine Is Yours: The rise of collaborative consumption*, New York: HarperBusiness.

Calder, A. (1995), *The Myth of the Blitz*, London: Pimlico.

Chappells, H. and E. Shove (2005), "Debating the future of comfort: Environmental sustainability, energy consumption, and the indoor environment," *Building Research and Information*, 33(1), 32–40.

Cohen, M. (2011), "Is the UK preparing for 'war'? Military metaphors, personal carbon allowances, and consumption rationing in historical perspective," *Climatic Change*, 104, 199–222.

Cohen, M. (2013), "The fall and decline of consumer society?" *Great Transition Initiative*, May. Available online at http://www.greattransition.org/images/GTI_publications/Cohen_The_Decline_and_Fall_of_Consumer_Society.pdf (accessed 18 August 2014).

Connelly, M. (2004), *We Can Take It! Britain and the memory of the Second World War*, New York: Pearson Longman.

Cox, S. (2013), *Any Way You Slice It: The past, present, and future of rationing*, New York: New Press.

Crawford, M. (2010), *Shop Class as Soulcraft: An inquiry into the value of work*, New York: Penguin.

Department of Environment, Food, and Rural Affairs (DEFRA) (2008), *Synthesis Report on the Findings from DEFRA's Pre-feasibility Study into Personal Carbon Trading*, London: DEFRA.

Dickinson, J. (2009), "The people paradox: Self-esteem striving, immortality ideologies, and human responses to climate change," *Ecology and Society*, 14(1).

Douglas, M., S. Watkins, D. Gorman, and M. Higgins (2011), 'Are cars the new tobacco?' *Journal of Public Health*, 233(2), 160–169.

Eames, M., T. Dixon, T. May, and M. Hunt (2013), "City futures: Exploring urban retrofit and sustainable transitions," *Building Research and Information*, 41(5), 504–516.

Fawcett, T. (2012), "Personal carbon trading: Is now the right time?" *Carbon Management*, 3(3), 283–291.

Fawcett, T., C. Bottrill, B. Boardman, and G. Lye (2007), *Trialling Personal Carbon Allowances*, Oxford: Oxford University, Environmental Change Institute.

Frye-Levine, L. (2012), "Sustainability through design science: Re-imagining option spaces beyond eco-efficiency," *Sustainable Development*, 20(3), 166–179.

Fujimori, S. and T. Masui (2011), "How dematerialization contributes to a low carbon society," *WIT Transactions on Ecology and the Environment*, 143, 315–326.

Gobble, M. and J. Euchner (2013), "The rise of the user-manufacturer," *Research on Technology Management*, 56(3), 64–67.

Gomi, K., K. Shimada, Y. Matsuoka, and M. Naito (2007), "Scenario study for a regional low-carbon society, *Sustainability Science*, 2(1), 121–131.

Grahn, M., E. Klampfl, M. Whalen, and T. Wallington (2013), "Sustainable mobility: Using a global energy model to inform vehicle technology choices in a decarbonized economy," *Sustainability*, 5(5), 1845–1862.

Hanlon, P., S. Carlisle, M. Hannah, A. Lyon, and D. Reilly (2011), "Learning our way into the future public health: A proposition," *Journal of Public Health*, 33(3), 335–342.

Hatch, M. (2013), *The Maker Movement Manifesto: Rules for innovation in the new world of crafters, hackers, and tinkerers*, New York: McGraw-Hill.

Hopkins, R. (2008), *The Transition Handbook: From oil dependency to local resilience*, White River Junction, VT: Chelsea Green.

House of Commons (2008), *Personal Carbon Trading: Fifth Report of Session 2007–08*, London: Stationery Office.

Howell, R. (2012), "Living with a carbon allowance: the experiences of carbon rationing action groups and implications for policy," *Energy Policy*, 41, 250–258.

Huesemannm, M. and J. Huesemann (2011), *Techno-Fix: Why technology won't save us or the environment*, Gabriola Island, BC: New Society.

Huxley, R., B. Lloyd, M. Goldacre, and H. Neil (2000), "Nutritional research in World War 2: The Oxford Nutrition Survey and its research potential 50 years later," *British Journal of Nutrition*, 84(2), 247–251.

Keirstead, J. (2008), "What changes, if any, would increased levels of low-carbon decentralized energy have on the built environment?" *Energy Policy*, 36(12), 4518–4521.

Kirveennummi, A., J. Mäkelä, and R. Saarimaa (2013), "Beating unsustainability with eating: Four alternative food-consumption scenarios," *Sustainability: Science, Practice, and Policy*, 9(2).

Laestadius, L., R. Neff, C. Barry, and S. Frattaroli (2013), "Meat consumption and climate change: The role of non-governmental organizations," *Climatic Change*, 120(1–2), 25–38.

Lang, T. and D. Barling (2013), "Nutrition and sustainability: An emerging food policy discourse, *Proceedings of the Nutrition Society*, 72(1), 1–12.

Larsen, K. and M. Höjer (2007), "Technological innovation and transformation perspectives in environmental futures studies of transport and mobility," *International Journal of Foresight and Innovation Policy*, 3(1), 95–115.

Lipschutz, R. (2012), "Getting out of CAR: Decarbonisation, climate change, and sustainable society," *International Journal of Sustainable Society*, 4(4), 336–356.

Lombardini, C. and L. Lankowski (2013), "Forced choice restriction in promoting sustainable food consumption: Intended and unintended effects of the mandatory vegetarian day in Helsinki schools," *Journal of Consumer Policy*, 36(2), 159–178.

Longmate, N. (2002), *How We Lived Then: A history of everyday life during the Second World War*, Revised Edition, London: Pimlico.

Meyer, A. (2000), *Contraction and Convergence: The global solution to climate change*, Dartington, UK: Green Books.

Miliband, D. (2006), *The Great Stink: Towards an environmental contract*, London: Audit Commission.

Mont, O., A. Neuvonen, and S. Lähteenoja (2014), "Sustainable lifestyles 2050: Stakeholder visions, emerging practices and future research," *Journal of Cleaner Production*, 63(15), 24–32.

Morozov, E. (2013), *The Save Everything, Click Here: The folly of technological solutionism*, New York: Public Affairs.

Neuvonen, A., T. Kashinen, J. Leppänen, S. Lähteenoja, R. Mokka, and M. Ritola (2014), "Low-carbon futures and sustainable lifestyles: A backcasting scenario approach," *Futures*, 58, 66–76.

Paterson, M. and J. Stripple (2010), "My space: Governing individuals' carbon emissions," *Environment and Planning D: Society and Space*, 28(2), 341–362.

Peters, G. (2008), "From production-based to consumption-based national emission inventories," *Ecological Economics*, 65(1), 13–23.

Powell, T. and T. Lenton (2012), "Future carbon dioxide removal via biomass energy constrained by agricultural efficiency and dietary trends, *Energy and Environmental Science*, 5(8), 8116–8133.

Scarborough, P., S. Allender, D. Clarke, K. Wickramasinghe, and M. Rayner (2012), "Modelling the health impact of environmentally sustainable dietary scenarios in the UK," *European Journal of Clinical Nutrition*, 66(6), 710–715.

Schandl, H. and G. Turner (2009), "The dematerialization potential of the Australian economy," *Journal of Industrial Ecology*, 13(6), 863–880.

Schor, J. (2011), *True Wealth: How and why millions of Americans are creating a time-rich, ecologically light, small-scale, high-satisfaction economy*, New York: Penguin.

Seyfang, G. and A. Haxeltine (2012), "Growing grassroots innovations: Exploring the role of community-based initiatives in governing sustainable energy transitions," *Environment and Planning C: Government and Policy*, 30(3), 381–400.

Sgouridis, S., P. Bonnefoy, and R. Hansman (2011), "Air transportation in a carbon constrained world: Long-term dynamics of policies and strategies for mitigating the carbon footprint of commercial aviation," *Transportation Research Part A: Policy and Practice*, 45(10), 1077–1091.

Stott, R. (2012), "Contraction and convergence: The best possible solution to the twin problems of climate change and inequity," *BMJ*, 344(e1765).

Vinnari, M. and P. Tapio (2009), "Future images of meat consumption in 2030," *Futures*, 41(5), 269–278.

Von Busch, O. (2013), "Molecular management: Protocols in the maker culture," *Creative Industries Journal*, 5(1–2), 55–68.

Waddock, S. and M. McIntosh (2011), "Business unusual: Corporate responsibility in a 2.0 world," *Business and Society Review*, 116(3), 303–330.

Williams, A. (2013), "Mobilities and sustainable tourism: Path-creating or path-dependent relationships," *Journal of Sustainable Tourism*, 21(4), 511–531.

Yip, C., G. Crane, and J. Karnon (2013), "Systematic review of reducing population meat consumption to reduce greenhouse gas emissions and obtain health benefits: Effectiveness and model assessments," *International Journal of Public Health*, 58(5), 683–693.

Zhang, Y. and Z. Zhu (2012), "Application study of the virtual reality reconstruction system link QTVR technology for sustainable tourism planning: The digital southern song palace," *International Journal of Digital Content Technology and Its Applications*, 6(16), 43–50.

Zweiniger-Bargielowska, I. (2000), *Austerity in Britain: Rationing, controls, and consumption, 1939–1955*. New York: Oxford University Press.

14

AFTERWORD

The waste that matters

Richard Wilk

This chapter is not in any way a summary of the book. Instead I have composed seven separate *provocations*, on the aspects of waste that I find most interesting and important. My approach is conditioned by a long-ago career as an archaeologist, and a more recent fascination with anthropological approaches to consumer culture as a historical and global phenomenon. In my view, the topic of waste is the most important connection between population, consumption and climate change, a key we desperately need in order to unlock the problem of sustainability.[1] To this end, I have chosen to write about the aspects of waste with which we most need to grapple. My approach is therefore more practical than it is overtly theoretical, though it is often hard to separate the two.

Waste is magic

Ask the average North American what happens to the waste from their sinks, toilets and washing machines after it goes down the sewage pipe. Many will not even know where that pipe is located, and fewer will know where their flush ends up. In most large towns and cities in the USA, the sewage pipes terminate at a treatment plant along with industrial wastes and sometimes storm water runoff, where it is held in a large tank to allow the solids to settle out. These solids are sometimes composted or used as fertilizer, though some are also landfilled or incinerated. The liquid fraction is put through a variety of biological treatments and filtration, and it may also be disinfected, before being pumped down into the ground where it pollutes well water, or more often, discharged into a nearby river, lake or sea. The next town or city downstream then takes its drinking water from that river, lake or aquifer purifies it and sends it out as tap water, which people drink and consider "clean" (though less than 10 percent is actually drunk or used in cooking, while each person flushes more than 110 liters down the toilet every day). Then it goes through the waste treatment cycle again, so a single drop of water in a long river may go through the process tens or even hundreds of times before reaching the ocean.

What if instead we just closed the loop, and took the effluent from the sewage treatment plant, and then disinfected and treated, sent it out to households again as tap water? Many closed loop processes have been invented and pilot plants have been built many times; they can produce water that is objectively cleaner and better-tasting than the average tap water, at much lower cost, which should make them very attractive in areas like California where water shortage is a perennial issue. But while they are scientifically and economically feasible, functioning "toilet to tap" water systems are very scarce; Singapore is the only major world city to get even a portion of its drinking water from sewage, and this only after a vast and expensive public relations program (see Taherzadeh and Rajendran this volume). Experts are always declaring that the era of recycled drinking water is right around the corner, but so far even drought-stricken cities have not been able to hook up the water taps to their septic outflow. Why is a city willing to drink water from a river, containing *someone else's* body waste, but not their own effluent? What is wrong with these people?[2]

Sanitation engineers don't understand something basic about waste that was discovered by Edward Burnett Tylor more than 150 years ago; that body waste is contaminated through principles of *magic* rather than chemistry. This pollution cannot be measured with any laboratory instrument, but it can be observed easily by anyone. Just ask a North American undergraduate student to drink a glass of water taken from a brand new, unused, and thoroughly disinfected toilet; or ask them to drink some of their own saliva; or try to get a Muslim student from South Asia to pick up a sandwich with the left hand. The principle of what is now called contagious magic says that contamination or other power can be passed through physical contact between people and/or objects. This principle of magic was asserted by Frazer (1911) and elaborated by Mary Douglas and Edmund Leach, who focused on basic cosmological divisions like culture/nature or animate/inanimate found in most human cultures (Douglas 1966; Leach 1958). They argued that substances or objects that are difficult to classify because they cross these cosmological boundaries are usually treated as powerful and dangerous, and a source of pollution, misfortune, and worse. Body waste, including hair and fingernails as well as feces and urine, are problematic because they are both part of the living body and inanimate and separable, not alive but part of life. This makes them powerful and contaminating substances in many cultures, often connected to supernatural power.

So why is it then acceptable to drink sewage water that has been diluted in a river? This brings to play another form of magic, the purifying power of nature. Elsewhere I have described the way bottled water companies play on the ambiguity of nature in modernist industrial societies (Wilk 2006a). Since the Romantic era in Europe, nature has been opposed to culture in a way that generates both purity and danger (Löfgren and Willim 2006; Löfgren 2002). Technology and medicine, in the form of purification plants and sanitary plastic bottles can cancel or cure the danger of nature, and nature can also take the dangerous products of human technologies and make them pure once again.³ The time that sewage spends in the river is enough to break the magical chain of pollution from body to feces. Why else would so many people be happy to spend their holidays bathing and playing in bodies of water that are often no more than diluted sewage? What technology pollutes, the natural can purify, and vice versa. The water has to be magically cleansed by going through the entire cycle before it can be consumed again as a magically pure substance.⁴

Even when human waste is a commodity to be bought and sold, it still has the magical power to pollute. In eighteenth-century Edo (now Tokyo), urine and feces were highly valued as fertilizer, for leather tanning, and other uses. A class of professional collectors bought human waste from each household, often bidding against one another, and they made regular collections, in a way that was usually invisible because privies were built against exterior walls (Hanley 1999; Francks 2009). In this case, the pollution did not inhere to the waste itself, but to the people who collected it, who were at the time known as *Eta*, a word meaning "an abundance of filth." This group also included workers in other impure professions including executioners, undertakers, workers in slaughter-houses, butchers, and tanners. Although Eta legal status as an outcast population was officially abolished in 1871, their descendants (now called *Burakumin*) continue to face discrimination. It is illegal today to hire someone to check on the ancestry of a prospective spouse, to make sure there is no genealogical "stain," but the practice persists, along with stereotypes about Burakumin "squalor, unemployment and criminality" (Wikipedia, http://en.wikipedia.org/wiki/Burakumin). The magical taint caused by association with waste is very difficult to erase, and can even be inherited.

Waste as class

As the Japanese example shows, dealing with waste is never a high-status occupation and it can carry a stigma (the Indian caste system is another good example). Professional handlers of waste often have the magical power to purify substances and put them back into social circulation, but in the process the impurity adheres to the agent. It seems almost a cultural universal that human waste is considered distasteful, but treating other kinds of waste as if they are as polluting as feces, vomit, or urine, is not. The connection between body wastes and other kinds of waste is purely metaphorical – there is no other connection between substances as diverse as leftover food from a meal, an old sandal washed up on a tropical beach, and a plume of gasoline from an old filling station that is now polluting the groundwater. The concept of waste is the only thing that holds them together.

Pierre Bourdieu made the now-familiar point that when people classify the things in the world, they also classify themselves (1984). So the classification of certain things as worthless waste is also an assertion of a particular kind of social status. In many urban societies, particularly in Europe and North America, class attitudes towards trash follow a common pattern I have previously called a "style sandwich" (Wilk 2006b: 173–5). A good culinary example would be wild game meat in early modern Europe; eaten by the poor (often illegally) through necessity, and by the rich as a demonstration of skill and power, but shunned by the middle class, who do not want to be associated with the poor.

Similarly, in the USA today, used clothes are bought in thrift stores out of necessity by the poor, and by people of high cultural capital looking for retro and fashionable items. In Bloomington Indiana, where I live, we have an annual "Trashion Fashion" Show, a fundraising event that promotes "sustainable design practices ... illuminates the importance of recognizing both environmental and social consequences of pollution and waste ... and showcases designs from upcycled clothing and discarded materials" (http://www.bloomingtontrashion.org/). The participants and the audience include artists and designers from university programs, people who are clearly well off, and/or have high cultural capital.

The style sandwich demonstrates the close relationship between consumption and production, and reminds us that conspicuous waste is often more important than conspicuous consumption as a form of social distinction. There are limits to how much a rich person can consume, even with the help of a large household and staff. Jacqueline Kennedy Onassis, for example, had trouble spending her annual clothing budget of $300,000; she often bought unusably large quantities of cosmetics and beauty supplies, and more clothing than she could ever wear, so her husband employed a staff of ten people whose job was to find ways to get rid of the excess in secret (Sparks 1970). While the public often finds the spectacle of rich people "throwing money around" fascinating and worthy of envy, they are also often disgusted by the waste. Not long ago, the targets of public ridicule were most often the nouveau riche, said to be clueless and tasteless wastrels, while the consumption of "old money" was hidden away behind gated walls, in exclusive communities, clubs, and settings where servants were the only common people allowed. But in recent times, for a variety of reasons, the border between new and old money has become less distinct.

Georges Bataille, the French surrealist philosopher, is still the only scholar to present a full theory of conspicuous waste. In *The Accursed Share*, he says that an expanding economic system produces surplus energy, so the real problem is not producing more, but is instead to find ways to get rid of that surplus in ways that do not tear society apart (Bataille 1991). The very goal of the economy is not to create, but to consume and produce waste. Earlier cultures, he thought, could get rid of surplus through sacrifice, feasting, and other public rituals of destruction like the potlatch, or they could build huge conspicuously useless objects like the pyramids. Under capitalism, though, sacrifice is forbidden as wasteful, so war on an unprecedented scale becomes the most common way to dissipate and eliminate surplus. Living through two world wars and the great depression makes this cynical view of the economy more understandable.

Bataille sees waste as a transcendent social fact, not just an indicator of how much is consumed, but the actual goal of labor, so that ownership and investment are mere distractions, or impossibilities. This is a good description of cultural settings where accumulation is forbidden, as in Sahlins' famous description of generalized reciprocity, where property itself is barely recognized in many hunting–gathering societies (Sahlins 1972). One could reverse this view to say that everything must be consumed so that no surplus exists, even if this means wasting food or destroying possessions. Paleolithic humans left sites behind where so many animals had been killed that only small amounts could be consumed. The anthology *Lilies of the Field* has a number of ethnographic cases of groups of people who resist any sort of accumulation, like the Veso of Madagascar who take pride in not knowing where the next day's meal will come from (Astuti 1999).

We also find this "ideology of waste as freedom" among the poorest and most exploited workers at the margins of the world system, among the male working cultures of offshore fishermen, plantation workers, trappers, cowboys, loggers, placer miners, and the mariners of the age of sail. As I have detailed elsewhere, in the "crew culture" of these mobile work gangs, men lived most of the time in the cash-free economy of camps and ships, subsisting on rations (Wilk 2014, 2006b). When they were paid off, they typically went to the nearest "sailortown" or "red light district" were they spent their money recklessly and quickly on prostitutes and drink, and lost the rest at gambling. Fighting, brawling, and public drunkenness often led them to end up in jail, and thence back to work.

Before or after the binge they often spoke of their pay as a "burden" or as something "hot" or "slippery." Their spending tended towards the spectacularly wasteful; buying drinks for everyone in a saloon, lighting cigars with burning bills, or buying expensive jewelry or clothing that ended up being pawned or stolen. All of these forms of enjoyment were highly social, building a man's status as a valuable member of his profession and crew, and generosity was an important part of that status, as in so many other cultures that anthropologists have studied. Binge spending kept these men stuck in a life of debt and both dangerous and poorly paid labor, yet they often described their binges as a kind of freedom. Once again, we see how waste can give a subjective experience of transcendence, and a way of creating social class identities.

Conspicuous thrift

My father was very conscious of social status, since his father was a self-made man who became an important man in Hollywood, and traveled widely in literary circles (Wilk 1975). When I was an adolescent, my father set out to pass along everything his father had taught him about consuming like a cultured gentleman. I was taught to taste single-malt whiskeys, choose well from a French menu, and buy the right kind of clothes. As Max explained, you went to London to a bespoke haberdashery where they knew your family; you chose the most durable finely woven wool cloth for your custom-made suits, and were fitted for a pair of hand-made shoes. He explained that these goods seemed ridiculously expensive when you bought them, but they lasted a lifetime. The same kind of thing went for hats, umbrellas, watches, pipes, and even socks, cologne, and a university education. Buy the best and your style will be timeless.

This was the code of "old money" in the twentieth century and it still exists in some forms today. The strategy is based on investing in long relationships with important people, deep cultural knowledge, and valuable things, so that possessions will always be of the kind that increase in value, like fine furniture, silver, classic cars and fine art. An expensive fountain pen is the kind of object you will pass down to your descendants, instead of buying cheap ballpoints that fall apart and get lost or discarded. Objects should acquire what McCracken calls "patina," the physical trace of time and stewardship (1988). In the long run, I was told, it was cheaper to invest your money in great things and top brands, not to demonstrate your wealth, but to save it. The implication was that only the nouveau riche built extravagant houses in outlandish or regimented styles, drove in flashy but uncomfortable cars, and bought big yachts. Contrary to what many people assume, thrift is a virtue of the rich.

Today that economic logic has been undercut by the proliferation and abundance of cheap things. As Campbell points out in this volume, it is often cheaper to go through several cheap throwaways, rather than spending money on durable top-of-the-line objects, especially when fashions move so quickly that many gadgets are obsolete before they are worn out. But there are still "durable" things like furnishings, art, watches, jewelry, and some kinds of clothes, where expensive higher quality still endures and lasts, while cheaper things fall apart or easily break. This is a matter of both the quality of materials and the skill with which they are combined. Think of the contrast between expensive cast iron or wooden toys, and cheap ones made of thin plastic.

Since my father's generation, a lot of people have become rich enough to buy the best enduring things, and then throw them away anyway, even if they are not broken or worn out (a practice that luxury designers deplore). At the same time, the poor end up paying more when it comes to things like housing. A badly built house or a prefab mobile home (which are as much as 18 percent of houses in some states) loses value over time, and mobile homes eventually become waste. Because they lack storage space and ready cash, poor people in the USA tend to buy smaller packages more frequently, which means they pay for (and discard) proportionately more packaging (Rathje and Murphy 2001). The thrift of the poor is often inconspicuous or even hidden from sight, invested in social relationships rather than securities or property.

Waste as time

I went to graduate school as an archaeologist, and had the good fortune to work as a teaching assistant for Michael Schiffer, who was trying to establish a more scientific and less culturally-weighted way of looking at human material culture over long periods of time (Schiffer 1976). One of the class assignments we developed was to send students out to map the many vacant lots in Tucson, looking for the material traces of behavior. We wrote a paper about the very surprising things they had found, which included piles of low-value building materials like sand and adobe blocks (Wilk and Schiffer 1979). People would slowly accumulate building materials on their lot until they had enough to start building a house, but it was clear from the condition of the piles that that magic moment never arrived for a lot of families. We called this in-between state a "storage-abandonment continuum," recognizing a very slow process as useful materials became useless waste, though they could be rescued or revived right up to the very last stages.

The point of this example is that the very status of something as "waste" depends on the kind of temporal framing. Thinking in archaeological time, Schiffer spoke of materials and objects entering the "systemic context" of human engagement through extraction, manufacture, and use. This context lasts as long as the objects remain engaged in human lives, and then they enter the "archaeological context" through "site formation processes." They may be purposely buried, lost, abandoned, or deliberately destroyed, all of which leave characteristic traces, and things remain in archaeological context until they are destroyed or recovered, ideally by archaeologists, but often by looters and careless builders.

Seen this way, artifacts and traces spend most of their lives in archaeological context, punctuated by brief periods of engagement with people. Also, from this long span perspective, we can imagine waste as a very short, transient phase in an object's life just before it enters the archaeological record, or we have to recognize that everything from the past that has not been recovered and put in museums, is waste, rubbish, garbage. Most archaeologists tend towards the latter view, that everything archaeologists find is the refuse of earlier times. And in this sense, waste is the most important and valuable thing people leave behind, the precious traces of how people actually live, as opposed to documents and monuments, which follow symbolic and political agendas.

We can also think of the life of an artifact with the tools offered by the authors of Appadurai's edited *The Social Life of Things*. Material culture can be a commodity, or a meaningful and inalienable kind of property; think of the difference between a new mobile home and Thomas Jefferson's home, Monticello. One is an anonymous duplicate with only market value, the other is a sacred relic that could not be bought or sold for any amount of money. During their "use life" (another archaeological term) an object can move back and forth between commodity and relic (or *gift* in the language of economic anthropologists). But as a general rule, the longer people own things, the more they share those things with people they love and care about, the more the object moves away from being a commodity, as it acquires value based on sentiment, morality, religion, and cosmology, developing its own singular identity. This helps make Miller's point that waste is not a problem caused by materialism, but by people not loving material things enough to want to keep them (see Miller 2009). A really materialist society would hardly throw anything away.

As a general rule of thumb, however, most objects only begin their lives as commodities, and the longer they are kept by an individual or group, the "stickier" they become through sentiment and memory. This may be why so many North American "hoarders" are filling their houses with piles of things they cannot bring themselves to throw away, objects stranded on their path along the storage-abandonment continuum, stuck to the people who love them (Arnold *et al.* 2012).

On a longer time scale, we can look at present municipal solid waste deposits in landfills as time capsules, or more likely, as a kind of saving for the future, an "unnatural resource." After all, today's fossil fuels are no more than the compressed and heated detritus of pre-human forms of life. As Taherzadeh and Rajendran predict (in this volume), someday people may look at our contemporary waste as a cache of valuable materials worth *mining*, and this may not be in the distant future. Already we are tapping landfills as sources of biogas to generate electricity and fuel incinerators. From this perspective, it is the people recycling garbage who are the worst wastrels, since they are depriving future generations of a time-deposit of energy and materials. We can even anticipate archaeologists and preservationists campaigning to save landfills, or a moral panic over the "coming garbage shortage."[5] At the time scale of geology, of course, even this perspective fades into insignificance. Counting in millions of years, our Anthropocene epoch will be a few meters of rock, full of "technofossils" and layers of "plastiglomerates" (Chen 2014).

Waste as place

Space is just as essential as time in defining and understanding waste, and there are whole regions designated as *wasteland* for various reasons. As several authors in this book have observed, global capitalism spatially separates waste from consumer goods in such a way that it is very hard to connect them, a phenomenon Princen calls "distancing" (2002). Most of the electricity and fossil fuels that power our households and vehicles are completely distanced from their sources; no watt-hour carries a stamp telling of its origins, and we only have a general idea of what kinds of resource mix is used by particular utilities to generate power (not to mention that only about 40 percent of the energy produced by burning the fuel is actually converted into electricity; the rest is "waste heat"). Petroleum certainly has "terroir" when it enters a refinery, since every oil deposit has its signature qualities, but the fuels and raw materials that emerge from the factory are anonymous and "de-sourced" just like the high fructose corn syrup that seems to be an inescapable ingredient of any meal in the USA. Corporations often find that distancing is an important marketing strategy, as Lilienfield and Rathje found out when they tried to account for all the costs of a few common artifacts (1998).

Even further back in the supply chain of even simple products are all the waste material and energy related to the extraction, transportation, and refining of raw materials, including the "virtual water" and "embodied carbon" involved at every stage of production, as well as the food, water, and other resources that sustain the human workforce required in all the stages of production from prospecting to retail

(Harriss and Shui 2010). Mining exotic minerals like the rare-earth elements may produce thousands of tons of waste (tailings), some of it highly toxic, for a single kilogram of the desired substance (Saleem 2014). Producing a single kilogram of gold requires, on average, 691,000 liters of water, 141 kg of cyanide, and 143 gigajoules of energy, emitting 11.5 tons of CO_2 (Mudd 2008).

The actual "overburden" of earth or rock that needs to be moved to access minerals beneath is yet another form of waste that is rarely included in a final accounting, because mine operators don't release this information. The material is sometimes reused later as "backfill" or to contour the new landscape left after mining, as in mountain-top removal coal mining in Appalachia, USA. Most types of modern mining also require vast amounts of water, which may be polluted and which often escapes into watercourses or to groundwater. Accounting for waste that occurs very far from industrial products, each one of which can incorporate hundreds of different materials from as many sources, we find that only about 3 percent of the total waste produced in the USA is municipal or household waste. The rest is industrial waste of many kinds. Max Liboiron argues that most municipal waste should also be seen as industrial waste that has been dumped on consumers, in the form of packaging and shoddy goods, as a way to save money. Households then have to sort and recycle this waste for nothing, so industries can buy it back at a low cost (Liboiron 2013).

This all challenges the maxim that garbage is *matter out of place*. The largest amounts of waste; as tailings, slag heaps, settling ponds, and overburden never move very far from their starting place. What makes this anonymous stuff into waste is the fact that humans have moved it to get to something else. It might be more accurate to say that garbage is matter that we *can see and/or smell*, but which we would rather not. We can add the corollary that the most dangerous waste is generally that which we cannot see, smell or taste, invisible poisons, heaps of toxic rock leaking into rivers and reefs, all far from the consumer. Much of the effort of reforming and reducing waste goes into making these distant problems visible and legible again (Dauvergne 2008; Crang *et al.* 2013). On the other hand, the high visibility of the relatively tiny amount of trash that litters streets or the plastic washing up on a beach gets people excited and motivates public action, again emphasizing that perceptibility is just as important as location in making things into waste.

Does waste really exist?

The examples of cloth loops described by Gustafsson *et al.* in Chapter 7, and the clothing recycling discussed by Ekström *et al.* in Chapter 9 make the important point that what a person or group see as waste can be a resource for someone else. Artists pick up different colored plastic from beaches to make sculptures, following a long tradition of beachcombing; and villages in Kenya specialize in carving animals from washed-up sandals. As demonstrated in Chapter 4 by Taherzadeh and Rajendran, defining waste is also subject to cultural differences, particularly in the way people classify and order things in language and cognitive categories. The Q'eqchi' rainforest farmers I lived with in 1980 did not have a general category of trash or waste; they reused almost everything including old bottles and broken pottery. The kinds of waste they did recognize included things like rice hulls, floor sweepings, and mud stuck to one's feet, and even the floor sweepings were sifted to get dust that was then mixed with water to plaster the floor.

The concept of waste also has a strong moral component too, multiplying the degree of subjectivity that is wrapped up in the category. If we make a simple list of all of the kinds of things that can be called waste, from an idle moment to a mountainous landscape, it should be clear that we are dealing with a folk-concept rather than an objective analytic category. This is why disciplines have to come up with their own narrow definition of waste, which is suitable for their own task, but inevitably leave out things that other scientists consider important in their work. The miscellaneous and heterogeneous things that fall into the category of waste are not always connected to one another; instead they form what Lakoff calls a "radial domain," a fuzzy set of things held together by their metaphorical connection to a prototypical object (Lakoff 1987). These prototypes, according to Lakoff, are often based in direct bodily experience, like floating or falling (Lakoff and Johnson 1999). The dirty, offensive and polluting nature of waste suggests that in the English-speaking world at least, the prototype for waste is fecal, so that all the things designated *waste* have some metaphorical relationship to shit. But the fact that the different kinds of waste have only an indirect relationship to each other means that each kind of waste has different potential for re-use or recycling, carry very different degrees of danger and hazard, and have different legal status. So one kind of industrial ash is easily recycled as an additive in road concrete, while another has to be expensively treated as hazardous waste (a current point of contention in states in the USA that burn coal for power).

Another related aspect of the subjectivity of waste is the way it elicits certain stereotypical forms of discourse, a recognizable genre we might call "trash talk," inspired by Nichter's discussion of "Fat Talk" among young American women (2001). As a speech genre, Trash Talk articulates the taken-for-granted nature of waste, what "everyone" understands, like the notion that all waste is noxious or ugly. Just from my own limited experience, I can identify five regular themes in Trash Talk.

1. Morality – be thrifty, don't throw away useful things, don't buy more than you need, clean up your own mess.
2. Group boundaries – *those* people leave their garbage everywhere, we would never waste money that way.
3. Health – waste is filthy and causes disease, pollutes water, land, air, is killing us.
4. Temporality – we used to be/have become frugal, respectful of nature, clean and healthy, now all is filth (or vice versa).

5. Crisis talk – if this keeps up we will drown in trash, die from pollution, run out of some resource, have no place left to dump.

Finding repeated speech genres does not in any way challenge the truthfulness or usefulness of any of these statements. But we do need to recognize when we are falling into easy pathways based on common knowledge. Part of our role as scholars is to keep questioning the common wisdom and finding new ways to frame old problems, so we do not end up repeating actions and programs that have failed in the past. If we are going to face the serious problems of the present and future, we need to be able to imagine, for example, that exposing children to dirty environments and polluted water may actually make them healthier in the long run. We might also be able to think more constructively about which kinds of waste are not serious hazards, even if they seem immoral or disgusting.

The waste that matters

The chapters in this book are concerned with the material refuse of industrial society, a problem that is threatening because of the vast amounts of refuse generated in satisfying the consumer demands of the richest fraction of the world's people. If, as presently anticipated, another two billion people are going to have enough money to join mass consumer culture in the next few decades, we will have to cope with staggering new amounts of household waste. The authors of these chapters are thinking their way forward into societies that produce less waste, even with the same or higher standard of living. But we are really at the very beginning of this process; we still live in a world dominated by an ideology of growth where consumption and discard are the most important realization of that ideology. So here are my final provocations.

Michael Maniates uses the term "individualization" for the common tendency in market capitalism to push social and systemic problems onto the individual (Maniates 2002). If we would all just take shorter showers and let our lawns die, we will solve the water shortage problem and save the whales. But in most places in the USA, industry and agriculture use more than 90 percent of the available fresh water, much of which is used inefficiently and wasted because of direct and indirect subsidies. No matter how much water domestic consumers save by taking shorter showers or letting their lawns die, it will make very little difference when it comes to water consumption as a whole. At a higher level, prices affected by international trade compacts or government regulation may have a much larger effect on waste production than millions of households changing their habits. This raises an issue of strategy as well in that it is probably going to be cheaper to convince a few bureaucrats or lawmakers to change the rules than it is to pay the cost of reaching millions of individuals.

The standard economic approach to reducing waste depends on a model where individuals make decisions. But most people are not autonomous individuals—they live in households with others, and they have extended networks of kin and close friends, and even animal companions who have their own needs and desires. A number of programs and agencies promoting sustainable living in the USA have found that disagreement in the household is a common obstacle to changing consumption habits and waste disposal. The question of who is going to carry out the compost and sort the trash may be more important than individual feelings about being green.

And if we do get people to change their habits and produce less waste, will this really reduce the amount of waste they consume, or just change the types of waste and where they end up? The phenomenon of "rebound effects" means that when people do something more efficiently and save money, these savings will be spent on other things. If the household starts buying used clothing, and they use the money they save for a driving holiday, we have just shifted the waste from one place to another, and we may have actually increased environmental damage. Even if they put the money into a bank, that bank is going to invest it in loans for projects as diverse as petrochemical factories, mortgages on new houses and consumer credit. Or they may invest in companies that are in the business of turning durables into consumables, as discussed by Campbell in Chapter 2.

These complications make Cohen's discussion of rationing in Chapter 13 particularly important. As long as we have large groups of rich people in our society, there is always going to be a fairness issue when it comes to paying a carbon tax or using price to regulate consumption or reduce waste. What incentive do poor people have to change their behavior when every day Donald Trump is throwing away more than they waste in a year? Why should I be composting my little dribble of vegetable waste while the hospital and grocery stores in my town are dumping vast amounts of edible food every day? Rationing is not the only way to promote fairness; other societies in history have used public shame, high taxation and sumptuary laws to restrain the spending of the richest. Violent revolutions, religious fundamentalism, ethnic strife, and regime changes have also been motivated by the visibility of vast inequalities.

Finally, at this particular moment I believe the highest priority should be given to identifying and addressing the kinds of waste that emit the most greenhouse gas into the atmosphere.[6] Who are the biggest wasters? This in itself is a complex task because there are so many sources, of such diverse nature that solutions will be complex. As a species, we might survive drastic climate change that raises sea levels by two or three meters by the end of this century, but it is not going to be a pleasant world to live in. We are simply not doing enough to curtail greenhouse gas pollution, and so far we are not seeing technological breakthroughs that might lead to rapid reductions in emissions or greater carbon capture in the environment. We run the real danger of wasting our entire planet and the time we could have spent saving it. The authors of this volume are clearing paths towards a more sustainable future, starting a conversation that I hope many more people will join in the immediate future.

Notes

1 My approach is also informed by the literature in material culture studies, and more specifically by the work of Elizabeth Shove (e.g. 2003) and David Evans (2012, 2014) as well as a long-term engagement with the work of Daniel Miller and many colleagues in the emerging field of Sustainable Consumption. My approach is also very much centered on the USA, which probably wastes more of everything than any other country.
2 Under the headline "Texas Plant will Turn Sewage into Drinking Water" AP reporter Angela Brown explained local skepticism with this anecdote; "In June, officials in Portland, Ore., sent 8 million gallons of treated drinking water down the drain after a man was caught on a security camera urinating into a reservoir. City leaders said they didn't want to distribute water laced, however infinitesimally, with urine" (August 11, 2011, http://news.yahoo.com/texas-plant-turn-sewage-drinking-water-205103974.html, read June 18, 2014.)
3 In my analysis of water bottles and labels, I found nature in the form of pictures of mountains, springs or streams, and technology and science echoed in the shapes of bottles, and the statements of contents on the labels.
4 All *drinking* water is by definition impure; people do not like the flat taste of distilled water, so many bottlers have to add minerals.
5 There is precedent for trash shortages; in the early modern industrial economy, rags and bones were valuable because the supply was limited, and both were vitally important materials before wood pulp and petrochemicals displaced them.
6 Note also that most of the statistics we have on the contribution of different sectors to total GHG emissions are rarely accurate, and are often provided by the very industries that are being audited.

References

Appadurai, Arjun (ed.) (1986), *The Social Life of Things*, Cambridge: Cambridge University Press.
Arnold, Jeanne E., Anthony P. Graesch, Enzo Ragazzini, and Elinor Ochs (2012), *Life at Home in the Twenty-First Century: 32 Families Open Their Doors*, Los Angeles: The Cotsen Institute of Archaeology Press.
Astuti, Rita (1999), "At the center of the market: A Vezo woman," in Sophie Day, Evthymios Papataxiarchis, and Michael Stewart (eds), *Lilies of the Field*, Boulder, CO: Westview Press, 83–95.
Bataille, G. (1991), *The Accursed Share*, trans. Robert Hurley, New York: Zone Books.
Bourdieu, Pierre (1984), *Distinction: A Social Critique of the Judgment of Taste*, Cambridge, MA: Harvard University Press.
Chen, Angus (2014), "Will the fossil record preserve your computer?" *Science*, 344(6190), 1325.
Crang, Mike, Alex Hughes, Nicky Gregson, Lucy Norris, and Farid Ahamed (2013), "Rethinking governance and value in commodity chains through global recycling networks," *Transactions of the Institute of British Geographers*, 38(1), 12–24.
Dauvergne, Peter (2008), *The Shadows of Consumption*, Boston: MIT Press.
Douglas, Mary (1966), *Purity and Danger: An Analysis of the Concepts of Pollution and Taboo*, London: Routledge and Kegan Paul.
Evans, D. (2014), *Food Waste: Home Consumption, Material Culture and Everyday Life*, London: Bloomsbury Press.
Evans, D. (2012), "Beyond the throwaway society: Ordinary domestic practice and a sociological approach to household food waste," *Sociology* 46(1), 43–58.
Francks, Penelope (2009), *The Japanese Consumer: An Alternative Economic History of Modern Japan*, Cambridge: Cambridge University Press.

Frazer, J. G. (1911), *The Golden Bough*, London: Macmillan and Company.
Hanley, Susan B. (1999), *Everyday Things in Premodern Japan: The Hidden Legacy of Material Culture*, Berkeley: University of California Press.
Harriss, Robert and Bin Shui (2010), "Consumption, not CO_2 emissions: Reframing perspectives on climate change and sustainability." *Environment* November–December 2010. Available online at http://www.environmentmagazine.org/Archives/Back%20Issues/November-December%202010/not-co2-emission-full.html (accessed 18 August 2014).
Lakoff, George (1987), *Women, Fire, and Dangerous Things: What Categories Reveal about the Mind*, Chicago: University of Chicago Press.
Lakoff, George and Mark Johnson (1999), *Philosophy in the Flesh*, New York: HarperCollins.
Leach, Edmund (1958), "Magical hair," *The Journal of the Royal Anthropological Institute of Great Britain and Ireland*, 88(2), 147–164.
Liboiron, Max (2013), "Waste as profit & alternative economies." Discard Studies blog. Available online at http://discardstudies.com/2013/07/09/waste-as-profit-alternative-economies/ (accessed 18 August 2014).
Lilienfeld, Robert and William Rathje (1998), *Use Less Stuff: Environmental Solutions for Who We Really Are*, New York: Ballantine.
Löfgren, Orvar (2002), *On Holiday: A history of vacationing*, Berkeley: University of California Press.
Löfgren, Orvar and Robert Willim (2006), *Magic, Culture and the New Economy*, Oxford; New York: Bloomsbury Academic.
Maniates, M. (2002), "Individualization: Plant a tree, buy a bike, save the world?," in T. Princen, M. Maniates, and K. Conca (eds), *Confronting Consumption*, Cambridge, MA: MIT Press, 43–66.
McCracken, Grant (1988), *Culture and Consumption*, Bloomington: Indiana University Press.
Miller, Daniel (2009), *The Comfort of Things*, Cambridge; Malden, MA: Polity.
Mudd, Gavin (2008), "Gold mining and sustainability: A critical reflection," *Encyclopedia of Earth*. Available online at http://www.eoearth.org/view/article/153053/ (accessed 18 August 2014).
Nichter, M. (2001), *Fat Talk: What girls and Their Parents Say about Dieting*, Cambridge, MA: Harvard University Press.
Princen, Thomas (2002), "Distancing: Consumption and the severing of feedback," in T. Princen, Michael Maniates, and Ken Conca (eds), *Confronting Consumption*, Cambridge, MA: MIT Press, 103–132.
Rathje, William and Cullen Murphy (2001), *Rubbish! The Archaeology of Garbage*, Tucson: University of Arizona Press.
Sahlins, Marshall (1972), *Stone Age Economics*, Chicago: Aldine.
Saleem, H. A. (2014), "Social and environmental impact of the rare earth industries," *Resources*, 3(1), 123–134.
Schiffer, Michael B. (1976), *Behavioral Archaeology*, New York: Academic Press.
Shove, Elizabeth (2003), *Comfort, Cleanliness and Convenience: The Social Organization of Normality*, London: Berg.
Sparks, Fred (1970), *The $20,000,000 Honey-Moon: Jacki and Ari's First Year*, New York: Dell Publishing.
Wilk, Max (1975), *Every Day's a Matinee: Memoirs Scribbled on a Dressing Room Door*, New York: Norton.
Wilk, R. (2014), "Poverty and excess in binge economies." *Economic Anthropology*, 1(1), 66–79.

Wilk, R. (2006a), "Bottled water, the pure commodity in the age of branding," *Journal of Consumer Culture*, 6(3), 303–325.

Wilk, R. (2006b), "Consumer culture and extractive industry on the margins of the world system," in John Brewer and Frank Trentmann (eds), *Consumer Cultures: Global Perspectives*, Oxford: Berg Publishers, 123–144.

Wilk, Richard and Michael Schiffer (1979), "The archaeology of vacant lots in Tucson, Arizona," *American Antiquity*, 44, 530–536.

INDEX

Actor-network theory 7, 90, 98, 102, 115
Aesthetics 57, 146, 161

Bartering 203–4
Borrowing 26, 201

Capitalism 53, 228, 232, 235
Car-sharing 7, 89, 92, 94–7, 199, 205–6
Charity 5, 26, 59, 81, 123, 125–7, 154,
  159–60, 162–3, 172, 188
Carbon emissions 55, 216–18, 236
Climate change 9, 53, 172, 214, 217–18,
  225, 236
Collaborative consumption 8, 166, 200,
  202–4, 206–9, 219
Collecting 4, 7, 79, 115, 118–20, 125, 127,
  162
Combustion 6, 37, 69, 72, 80–1
Commodification 24–5
Commodity 180, 201, 227, 231
Community 6, 104, 116, 127, 168, 201,
  203, 206, 209, 216–17, 220
Composting 7, 61, 72, 74–5, 79, 80, 81, 89,
  102–3, 105–7, 109, 111, 141, 170, 236
Conspicuous 137, 228–9
Consumer agency 7, 93, 94, 102, 111
Consumer culture 56, 111–12, 135, 150–1,
  155
Corporate social responsibility 8, 118, 169
Corruption 69, 77–8, 82
Craze 44, 45

Democratic 52, 218
Decommodification 25
Decomposition 61, 104, 106
Digital revolution 209
Distinction 2, 228
Downshifting 54
Durability 155–6, 162–3, 189

Eco-fashion 43
Eco-taxation 89
Embeddedness 185, 200–201, 208,
Enrollment 107, 124–25
Environmental consumer socialization, 8,
  150, 153, 163
Environmentalist 33, 41, 153–5, 157–62,
  176
Ethical 122, 168–9, 171–2, 178–9, 189, 208
Ethics 4, 56
Everyday 2, 4, 7, 15, 21–2, 53–8, 60, 113,
  134, 137, 146, 179, 215, 219, 220
Exclusion 61
European waste directive 3, 58, 88

Fads 44
Fashion 1, 4, 5, 24, 26, 30, 36, 38–41, 43–4,
  116–17, 125, 128, 153–7, 161–2, 167–9,
  171–2, 175–80, 187–9, 193–4, 196, 228,
  230
Fashion cycle 5, 24, 187–8
Fashion-driven 36, 38
Fast fashion 5, 39, 41, 117, 128, 153–5, 161

Wilk, R. (2006a), "Bottled water, the pure commodity in the age of branding," *Journal of Consumer Culture*, 6(3), 303–325.
Wilk, R. (2006b), "Consumer culture and extractive industry on the margins of the world system," in John Brewer and Frank Trentmann (eds), *Consumer Cultures: Global Perspectives*, Oxford: Berg Publishers, 123–144.
Wilk, Richard and Michael Schiffer (1979), "The archaeology of vacant lots in Tucson, Arizona," *American Antiquity*, 44, 530–536.

INDEX

Actor-network theory 7, 90, 98, 102, 115
Aesthetics 57, 146, 161

Bartering 203–4
Borrowing 26, 201

Capitalism 53, 228, 232, 235
Car-sharing 7, 89, 92, 94–7, 199, 205–6
Charity 5, 26, 59, 81, 123, 125–7, 154, 159–60, 162–3, 172, 188
Carbon emissions 55, 216–18, 236
Climate change 9, 53, 172, 214, 217–18, 225, 236
Collaborative consumption 8, 166, 200, 202–4, 206–9, 219
Collecting 4, 7, 79, 115, 118–20, 125, 127, 162
Combustion 6, 37, 69, 72, 80–1
Commodification 24–5
Commodity 180, 201, 227, 231
Community 6, 104, 116, 127, 168, 201, 203, 206, 209, 216–17, 220
Composting 7, 61, 72, 74–5, 79, 80, 81, 89, 102–3, 105–7, 109, 111, 141, 170, 236
Conspicuous 137, 228–9
Consumer agency 7, 93, 94, 102, 111
Consumer culture 56, 111–12, 135, 150–1, 155
Corporate social responsibility 8, 118, 169
Corruption 69, 77–8, 82
Craze 44, 45

Democratic 52, 218
Decommodification 25
Decomposition 61, 104, 106
Digital revolution 209
Distinction 2, 228
Downshifting 54
Durability 155–6, 162–3, 189

Eco-fashion 43
Eco-taxation 89
Embeddedness 185, 200–201, 208,
Enrollment 107, 124–25
Environmental consumer socialization, 8, 150, 153, 163
Environmentalist 33, 41, 153–5, 157–62, 176
Ethical 122, 168–9, 171–2, 178–9, 189, 208
Ethics 4, 56
Everyday 2, 4, 7, 15, 21–2, 53–8, 60, 113, 134, 137, 146, 179, 215, 219, 220
Exclusion 61
European waste directive 3, 58, 88

Fads 44
Fashion 1, 4, 5, 24, 26, 30, 36, 38–41, 43–4, 116–17, 125, 128, 153–7, 161–2, 167–9, 171–2, 175–80, 187–9, 193–4, 196, 228, 230
Fashion cycle 5, 24, 187–8
Fashion-driven 36, 38
Fast fashion 5, 39, 41, 117, 128, 153–5, 161

Food labeling 135–6
Frugal 57, 136, 153, 234
Frugality 54, 134, 136

Generation 5, 8, 26, 30–1, 36–7, 40–1, 45, 136, 150–1, 153–63, 208, 230, 232
Gift giving 200–1, 203–4
Global warming 209
Government 1, 6, 32, 34, 37, 42, 45, 58, 67, 72, 76–9, 104–5, 109, 168–9, 173, 176–7, 216–17, 235
Greenhouse gas emissions 88, 214–15, 217–18, 236
Greenhouse gas pollution 236
Gross Domestic Product (GDP) 3, 69, 70, 74, 82, 84
Guilt 2, 15, 18, 20, 22, 26, 54, 109, 175

Habits 2, 6, 13, 15, 22–4, 53–6, 58, 97, 127, 137, 235–6
Health 32, 45, 55–6, 58, 67, 77, 88, 136, 219, 221, 234
Healthy 16, 136–7, 172, 234–5
Hedonistic 137, 161, 189
History 27, 33, 58, 103, 106, 134, 201, 218, 236
Hoarders 232
Hyper consumption 6, 29, 41

Identity 1, 54, 56, 60, 90, 93, 151, 155, 161, 189, 231
Incineration 3, 70, 72–3, 88, 98, 116, 125, 127
Inclusion 61
Inconspicuous 230
Individualism 6, 134, 138
Individualization 26, 235
Inequality 4
Industrial Network Approach 188
Infrastructure 78–80, 84, 97, 103–4, 111–12
Innovation 36, 38, 41, 45, 53, 57–8, 95–6, 138, 219
Intangible 189, 201, 207

Legislation 33, 36, 45, 71, 77, 80, 96, 202, 207
Legislators 168
Landfill(s) 3, 35, 59, 67, 69, 71–7, 80–2, 88, 92–3, 105–9, 112, 116, 171–3, 187, 225, 232

Leftovers 8, 60, 134–6, 140–6
Lending 201, 203, 206–7
Libraries 203, 208; clothing libraries 203; tool libraries 203; toy libraries 203
Lifespan 24, 35, 37, 162
Lifestyles 53, 58, 172, 216, 219, 220–1
Long-lasting 33–4, 170
Low-energy 42

Macro 6, 53, 55, 58, 219
Magic 16, 225–7, 231
Meso 6, 53, 58
Marketplace exchange 201, 203–4
Mass consumer culture 235
Materialism 103, 231
Materiality 6, 103
Material culture 2, 230–1
Memorabilia 19
Mending 153, 158, 162
Micro 6, 16, 53–4, 58, 78, 219
Minimalist 5, 13, 22, 27
Mobilization 7, 107, 111, 124, 127–8
Moral 15, 16, 21, 25–6, 52, 58, 60–1, 134, 136, 139, 146–7, 232, 234
Moral economies 26
Morality 60, 231, 234
Movement 14–16, 20, 33, 35, 43, 104, 167, 216, 220
Multiple-use 32–3
Mundane 4, 26
Municipal waste

NGO 76, 82, 105, 116, 119–22, 126–7, 166, 169, 173, 175–9
Negotiation 24–6, 126
Nouveau Riche 228, 230
Novel 6, 7, 29–30, 34, 36–7, 42, 43–5, 78, 194, 219–21

Obsolescence 33–5, 38, 40
Ontological 55, 60, 102
Over-consumption 3, 4, 5, 22, 30, 33, 41, 135–7, 179, 196
Outsourcing 25

Performativity 55, 135
Political agenda 3, 231
Possessions 2, 16, 18, 199, 209, 229–30
Post-consumer 5, 58–9, 188, 215, 219–20
Post-ownership 8, 9, 199

Practice theory 8, 59, 147
Primary socialization 8, 152, 154–9, 163
Privatisation 76
Producer responsibility 5, 71
Prosperity 3
Public-private partnerships 82

Resistance 45, 80, 104, 106
Rational 27, 56–7
Rationalization 56
Rationing 9, 158, 214–21, 236
Rebound effects 236
Recycle 1, 2–4, 8, 14–15, 23–4, 26, 36, 55–7, 59, 61, 69, 71–3, 81–2, 93, 97, 108, 110, 118, 150–1, 153, 159–60, 162–3, 170–3, 179, 188, 214, 226, 233
Recycling stations 4, 26, 79
Reductionist 29
Reflexivity 55, 57, 103
Relationship(s) 4, 25, 29, 91, 95–6, 103, 124, 138, 140–2, 144, 147, 152, 163, 189, 206, 228, 230, 234
Religion 69, 82, 231
Replacement 30–3, 35–7, 41–2, 154
Reuse 1–4, 8, 14, 36, 55–7, 59, 61, 69, 75, 82, 116, 118–19, 150–3, 159–60, 162–3, 166, 169–72, 177, 179, 188, 208, 214, 233–4
Risk 3–4, 26, 52, 54, 56–7, 109, 137, 174, 194
Romanticism 56–7
Routines 6, 13, 15, 23–4, 26, 54–8, 80, 137, 189–90, 194–5, 220–1

Salvation 57–8
Scavengers 74–5, 79
Second-hand 5, 31, 35–6, 43, 81, 91, 98, 120, 127, 154–7, 160–3, 188
Secondary socialization 8, 152, 156–7, 159, 162–3
Semiotic signs 14
Sharing 7, 8, 9, 25–6, 58, 89, 92–7, 166, 199–209
Sharing enterprises 207
Shopping 7, 21, 31, 33, 40, 54, 69, 98, 116, 119, 122–3, 125, 127–8, 154–7
Short-term 203–5, 207

Slow fashion 177
Social class 136–7, 154, 161, 227–9,
Solidarity 25, 53
Status 2–3, 9, 53–4, 67, 137, 151, 215, 227, 229, 231, 234
Storage 4, 17, 21, 122, 137, 141, 199, 209, 216, 230–32
Stuff 2, 6, 13–18, 21–6, 31–2, 37, 159, 166, 207, 233
Suburban 209, 221
Sumptuary laws 236
Surveillance 54
Swapping 26, 160–61, 177, 204
Symbolic consumption 1, 150, 153, 155
Symbolic messages 14

Tangible 189, 201, 207, 208
Taxation 236
Taxes 6, 61
Temporal 18, 30, 56, 135, 138, 140, 146–7, 231
Temporality 234
Temporalities 18, 24, 135, 137
Trading 45, 124, 154, 160, 202–4, 216, 218
Trajectories 16, 18, 24
Transformation 2, 14, 19, 54, 57, 103–4, 138, 167
Translation 90, 91, 97, 102–4, 106, 111–12, 124

Urban 7, 52, 58, 75, 95, 98, 102, 109, 151, 209, 221, 227
Urbanization 219
Utility-scale provisioning 219

Value creation 93, 219
Vintage 5, 43, 118
Virtual reality 219
Vulnerable 52, 116

Warranty 35, 174
Waste governance 98–9
Waste hierarchy 3, 5, 9, 58, 69, 88, 92, 97–8, 214
Waste separation 69, 71, 75, 79
Waste-sorting machines 79
Welfare 45, 67, 98, 116, 151